PROCESS TECHNOLOGY

edited by

Elisabetta Di Nitto
Alfonso Fuggetta
Politecnico di Milano and CEFRIEL

A Special Issue of
AUTOMATED SOFTWARE ENGINEERING
An International Journal
Volume 5, No. 1 (1998)

SPRINGER SCIENCE+BUSINESS MEDIA, LLC

ISBN 978-1-4613-7484-8 ISBN 978-1-4615-5441-7 (eBook)
DOI 10.1007/978-1-4615-5441-7

Library of Congress Cataloging-in-Publication Data

A C.I.P. Catalogue record for this book is available
from the Library of Congress.

Printed on acid-free paper.

AUTOMATED SOFTWARE ENGINEERING

An International Journal

Volume 5, No. 1, January 1998

Special Issue: "Process Technology"
Guest Editors: Elisabetta Di Nitto and Alfonso Fuggetta

Distributors for North America:
Kluwer Academic Publishers
101 Philip Drive
Assinippi Park
Norwell, Massachusetts 02061 USA

Distributors for all other countries:
Kluwer Academic Publishers Group
Distribution Centre
Post Office Box 322
3300 AH Dordrecht, THE NETHERLANDS

Library of Congress Cataloging-in-Publication Data

A C.I.P. Catalogue record for this book is available
from the Library of Congress.

Printed on acid-free paper.

Automated Software Engineering 5, 5–6 (1998)

Special Issue on Process Technology

ELISABETTA DI NITTO dinitto@elet.polimi.it
ALFONSO FUGGETTA fuggetta@elet.polimi.it
Politecnico di Milano and CEFRIEL, Via Fucini, 2, 20133 Milano, Italy

Recent years have seen a significant development of methods, techniques, and technologies supporting and coordinating different kinds of process-based activities. These contributions have emerged within the context of different disciplines. Software engineering has originated a variety of process-centered environments and process modeling languages aimed at describing and supporting the work of software developers. Within the context of information system development, a variety of technologies and methods have emerged to describe the workflow adopted by organizations to achieve their business goals. Finally, tools commonly known as CSCW environments have been developed to support, in general, cooperative activities.

As a result, a new discipline is emerging, whose goal is to develop technologies and methodologies to support human-centered cooperative activities. Even if the title is certainly not completely appropriate, we may call it *process technology*. The technologies that have been developed so far share several similarities and open issues. They are typically based on the notion of "process model"; i.e., an explicit description of the activities, procedures, and artifacts that constitute the process followed to accomplish the goals of the organization. The development and validation of process models is supported by specific editing and analysis tools. Once a process model is considered accurate and complete, it can be "enacted"; i.e., interpreted by a tool (often called a process engine) which is able to enforce, support, and track the achievement of process goals and objectives. Process engines guide and monitor the activities executed by the humans involved in the process, and automatically perform the process steps that do not require human intervention.

Process technology needs to be integrated with application-specific tools that automate parts of the process being supported. For instance, in software development a variety of tools are used such as compilers, editors, and CASE tools. Similar issues are addressed when integrating a workflow technology with business-specific tools. In general, process technology must guarantee the ability of operating business specific tools according to the guidelines expressed in the process model.

Another important issue that is relevant to all these technologies is their ability to tolerate and manage inconsistencies, deviation, and process evolution. This is needed to take into account the intrinsic dynamics of the organizations where process technology will be used.

In summary, there is a body of knowledge, technique, methods, and technology that is emerging as a result of the study of different kinds of processes. However, even if there are significant results, both in terms of research contributions and industrial practice, we have reached a very critical stage in the development and deployment of this technology:

- Existing systems and techniques have emerged from different research domains, with little (if any) cross-fertilization of ideas and experiences. We miss a common understanding of the commonalties and differences among the existing approaches.
- We are experiencing a significant number of problems in adopting this kind of technology in real industrial and business settings.
- There are several commercial products in the market (especially workflow management systems), but we miss objective criteria to compare and assess them.
- The development of new telecommunication networks and infrastructure is changing the scenario where this technology is meant to operate.

One of the most critical problems is the lack of cross-fertilization among the different areas studying process technology. This special issue is just an initial, and still preliminary, attempt to create a shared vision of the problems and solutions related to process technology. The goal is certainly ambitious and there is a long way to go before a common background on process technology can be established, building on the experiences and contributions of different communities and cultures. Still, this is a promising beginning.

The special issue includes four papers. The first paper by Deiters and Gruhn discusses experiences and issues in using process technology in real business processes. The second and third papers by Grundy and Hosking, and by Dami, Estublier, and Amiour, respectively, present two new graphical languages to describe processes. Finally, the paper by Ben-Shaul and Kaiser discusses the issue related to distribution of processes and integration of multiple, and possible heterogeneous, process-centered environments.

We are pleased to offer the reader this selection of research contributions on process technology. We hope they can be instrumental for the development and further consolidation of the discipline.

Acknowledgments

We wish to thank the Editors-in-chief of the Automated Software Engineering Journal for the opportunity they gave us to edit this special issue. We wish also to acknowledge the hard work and dedication of the referees who did the critical job of evaluating the papers:

Sergio Bandinelli, Naser Barghouti, Israel Ben-Shaul, Antonio Carzaniga, Fabio Casati, Paolo Ciancarini, Reidar Conradi, Gianpaolo Cugola, Giorgio De Michelis, Flavio De Paoli, Wolfgang Emmerich, Jacky Estublier, Christer Fernstrom, Pankaj Garg, Volker Gruhn, George Heinemann, Medhi Jazayeri, Luigi Lavazza, Takuya Katayama, Karen Huff, Carlo Montangero, Barbara Pernici, Mauro Pezzè, Giuseppe Pozzi, Dieter Rombach, Richard Taylor, Giuseppe Valetto, Brian Warboys and Stefan Wolf.

Automated Software Engineering 5, 7–25 (1998)

Process Management in Practice Applying the FUNSOFT Net Approach to Large-Scale Processes

WOLFGANG DEITERS
FhG-ISST, Joseph-von-Fraunhofer-Str. 20, 44227 Dortmund, Germany

VOLKER GRUHN
University of Dortmund, Computer Science, 44221 Dortmund, Germany

Abstract. Management of business and software processes are areas of increasing interest, which evolved nearly independently from each other. In this article we present an approach to process management that has been applied to business and software processes and which, thereby, enabled cross-fertilization between both areas. The goal of this article is to report lessons learned in industrial as well as academic business and software process management projects.

Keywords: process management, software processes, business processes, FUNSOFT nets

1. Introduction

In the last decades many people have concentrated on techniques for achieving high quality software products. This led to the finding that properly managed software processes are one key factor for successful software development (Schäfer, 1995). It has been identified that software processes are one-of-a kind, i.e., no generic lifecycle model could be defined that encompasses sufficient detail to describe project-specific requirements for software processes. Therefore, techniques for defining individual processes, and for building software systems orchestrating the defined process have been defined. In the software engineering community, process centered software engineering environments (Bandinelli et al., 1994; Engels and Groenewegen, 1994; Junkermann et al., 1994) have been developed. In these environments data integration is implemented by repositories and control integration is based on software process models that are enacted by process engines.

In the fields of applied computer science and organization science, management of business processes (also called *workflow management*) has developed as an area of growing importance (Warboys, 1994) almost independent of software process management. Business process management (often called 'business process (re-)engineering' (Hammer and Champy, 1993)) is about modeling, analysis and enactment of any business processes. A business process is defined here as a logically connected set of activities serving a particular business purpose. Many tools for the support of business process management are not primarily dealing with the comfortable modeling but with the enactment of real business processes (Hales and Lavery, 1991).

A mutual stimulation of software process management and business process management seems to be promising, because in principle both are reaching for the same goal (the management of processes) while focusing on different aspects (modeling comfort versus most far reaching automated execution).

The aim of process management[1] is to (re)engineer processes in order to make them as effective and efficient as possible (Hammer and Champy, 1993; Warboys, 1994; Totland and Conradi, 1995). Therefore, a process should be **defined** and **analyzed** properly in order to obtain a means for discussions about the process and for process optimization and resource planning. An analyzed process model can be **enacted** in order to govern a process.

Our experience is that processes are not isolated entities, but they are inter-related. They have interfaces with each other and they exchange information. Delays in one process impact other processes. In a strict sense, all activities of related processes could be described as just one process model, but it turned out useful to understand complex process as being composed of simpler processes. This enables us to benefit from the expertise of various process owners, who can identify *their* process. As soon as a complex processes consists of simpler processes with different process owners, we call this process a large-scale process. Even though many process management experiments have been carried out (Bunje et al., 1996; Emmerich et al., 1996), only very few actually deal with clusters of related processes.

In this article we describe the FUNSOFT net (Deiters and Gruhn, 1994) approach to process management. This approach is the result of academic research work (Deiters, 1993; Gruhn, 1991) and research work in the international cooperation projects ESF (Eureka Software Factory) (Adomeit et al., 1994) and ALF (Accueil de Logiciel Futur) (Derniame and Gruhn, 1994). A commercial implementation of the FUNSOFT net approach (called LEU) is available.

The FUNSOFT net approach has been applied to both academic and industrial projects. The benefits of doing so is that we clearly identified a gap between research issues and real-world problems. While some of the research issues are really challenging and ambitious (e.g., on-the-fly modifications of process models) they do not focus on the most urgent real-world problems. On the other hand industrial projects reveal process management problems (e.g., process model version control and transaction concepts) which have not been identified as long as process management was applied to academic projects only. Thus, by applying the FUNSOFT net approach in academic *and* industrial projects, we enable cross-fertilization between the different process areas mentioned. This helped to refocus our current research work. The need for doing this is underpinned by Finkelstein et al. (1994) and Lonchamp (1994) where it is reported that only little experience in applying software process modeling to real-world processes is available.

The remainder of this article is structured as follows: In the following section we describe in detail how the different process management activities (modeling, analysis, enaction) look like in our approach. Section 3, then, describes two large process management projects. Section 4 summarizes the lessons learned, i.e., which *new* requirements were raised in the projects discussed and which extensions of our approach have been identified. Finally, Section 5 concludes this article by pointing out how valuable the combination of academic and industrial experience has been.

2. Process management based on FUNSOFT nets

In this section we describe basic features of the FUNSOFT net approach to process management, which are used in both projects discussed later. Further details of this approach are discussed in relation with the project examples (compare Section 3).

2.1. *Process modeling*

In the FUNSOFT net approach, process models consist of three basic building blocks:

- **Object models** are used to describe the structure of objects (and their relationships), which are manipulated within a process. In the FUNSOFT net approach object models are described by means of extended entity-relationship diagrams (Gruhn et al., 1995). A scheme of an object model is shown in the bottom part of figure 1[2]. Object types can either be of a predefined format (postscript, WordPerfect, etc.) or they can be structured. In case of structured object types, attributes have to be defined.
- **Activity models** are used to define activities to be executed in a process. Activity models are described by FUNSOFT nets (Deiters and Gruhn, 1994). FUNSOFT nets are high level Petri nets (Ghezzi, 1991), whose semantics is defined in terms of Predicate/Transition nets (Genrich, 1987). A scheme of a FUNSOFT net is sketched in the central part of figure 1. Activities to be carried out in processes are described by agencies (represented as rectangles). Agencies in FUNSOFT nets can be fired (i.e., the corresponding activity can be executed) as soon as all required input objects are available. Objects produced and manipulated by activities are stored in channels (represented as

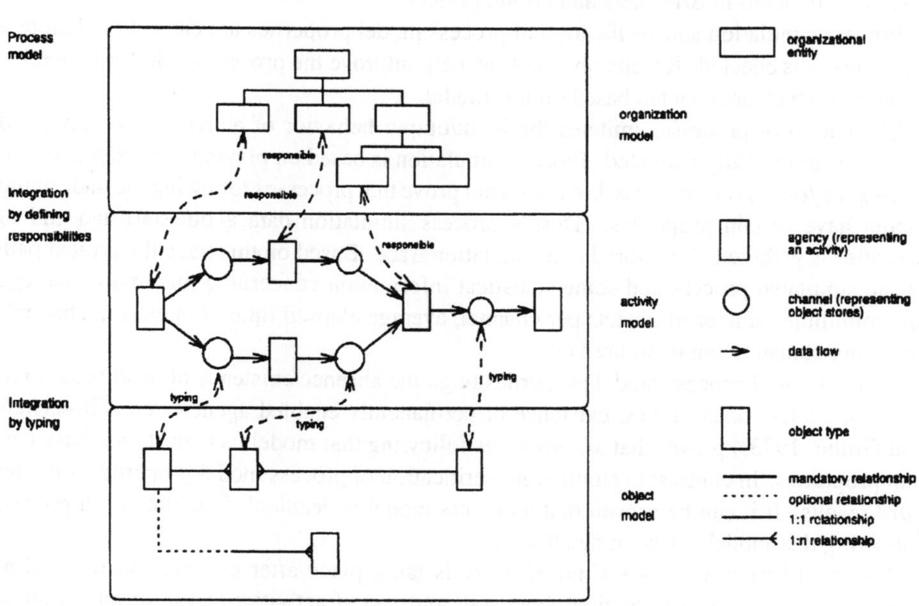

Figure 1. Integration of organization models, activity models and object models.

circles). FUNSOFT nets not only contain a definition of activities and their parameterization, but also an order of activities. They allow one to define that activities have to be carried out sequentially, concurrently or alternatively. The order is based on preconditions which have to be fulfilled before an activity can be executed.

- **Organization models** are used to define which organizational entities are involved in a process. The top part of figure 1 sketches an organization model. Each organizational entity is associated with a set of permissions for the execution of activities and for the manipulation of objects of certain types. A person assigned to an organizational entity *inherits* the permissions from that entity.

Once these aspects of processes have been modeled, it is necessary to integrate them. Integration of object models, activity models, and organization models means to define:

- which channels of FUNSOFT nets are typed by which object types identified in object models (indicated by the arrows annotated with *typing* in figure 1),
- which organizational entities are responsible for which activities (indicated by the arrows annotated with *responsible* in figure 1).

2.2. *Process model analysis*

Process model analysis helps to avoid that erroneous or inefficient process models are enacted. Process model analysis in the FUNSOFT net approach exploits the formal semantics of FUNSOFT nets. It covers process simulation, verification of process model properties, and evaluation of already finished processes. All three branches of process analysis are discussed in detail in Broeckers and Gruhn (1993).

Process simulation and verification of process model properties are carried out before a real process is enacted. Results obtained can help improve the process model and, thereby, processes which are enacted based on this model.

Simulation of processes imitates the input/output behavior of activities. A simulated process is graphically animated. Process simulation is based on playing the *token game*. It helps get a *feeling* for a process, but it does not prove that processes following the underlying model have certain properties. During process simulation data about start and end of activities is gathered. It is stored in a simulation trace. Based on this trace the critical path of the simulated process and some statistical information concerning maximum, average and minimum number of objects per channel, average elapsed time of objects in channels and other statistics can be figured out.

Verification of process model properties (e.g., the absence/existence of deadlocks, existence of useless object types, existence of permanently enabled agencies etc. (Broeckers and Gruhn, 1993)) proves that all processes following that model have or do not have certain properties. In contrast to simulation, verification of process model properties provides *hard* results. If it can be proven that a process model is deadlock-free, then each process following that model is free of deadlocks.

The third branch of process model analysis takes place after process enaction. During process enaction, information about start and end of activities and information about modifications of process models are recorded in enaction traces. Enaction traces are completed when enacted processes are finished. These traces resemble simulation traces, but they contain richer information, because modifications are recorded, and the start and end times of activities are real values. Moreover, it is recorded which process participant is involved in which activities. Enaction traces can be used to improve the underlying process model. These improvement affect future processes following that model.

2.3. Process enaction

The final goal of process management is process enaction. To enact a process means to ensure that a process behaves as specified in the process model. Activities which can be carried out without human interaction are started as soon as all their input parameters are available. Thus, process enaction frees process participants from menial activities by starting routine activities which do not require human intervention automatically. Process participants can concentrate on activities in which their creativity and problem-solving capabilities are needed. During process enaction all activities which could be carried out are identified and—provided they require human participation—offered to those process participants who have the appropriate permission. To offer an activity to someone means to write it to his/her agenda[3]. In addition, the documents and objects to be processed are made available. Once a participant chooses an activity from his/her agenda, the corresponding activity is started. This could, for example, mean to start a dialogue in which some information is to be added or manipulated, to start a text processing tool, or to start any other software. When an activity is started, the activity is deleted from all agendas. Once an activity is finished, the FUNSOFT net enaction component calculates the next process state and identifies which further activities could be executed. The agendas of all process participants are updated correspondingly.

Figure 2 sketches the relationship between FUNSOFT nets, agendas and the start of activities. On the top we recognize a small FUNSOFT net. All activities which could be executed and which require human interaction are put into the agenda of the process participants who have the appropriate roles. In the center of the figure we recognize the agenda of a process participant who is responsible for identifying apartments concerned by notices to quit and to carry out database searches. As soon as these activities are selected from the agenda, the associated tool or dialogue is started at the terminal of the process participant who selected it (bottom of figure 2). Automatic activities (e.g., *archive*) are started as soon as all their input parameters are available.

Summing up, process enaction means to ensure a flow of information and a coordination of process participants that corresponds to the process model.

3. Application of the FUNSOFT net approach to process management

In this section we describe two of the projects in which the FUNSOFT net approach has been used. The project described in Section 3.1 focuses on analyzing and optimizing a software process for maintaining a safety critical software system. The project discussed in Section 3.2 describes the management of business processes for the housing construction and

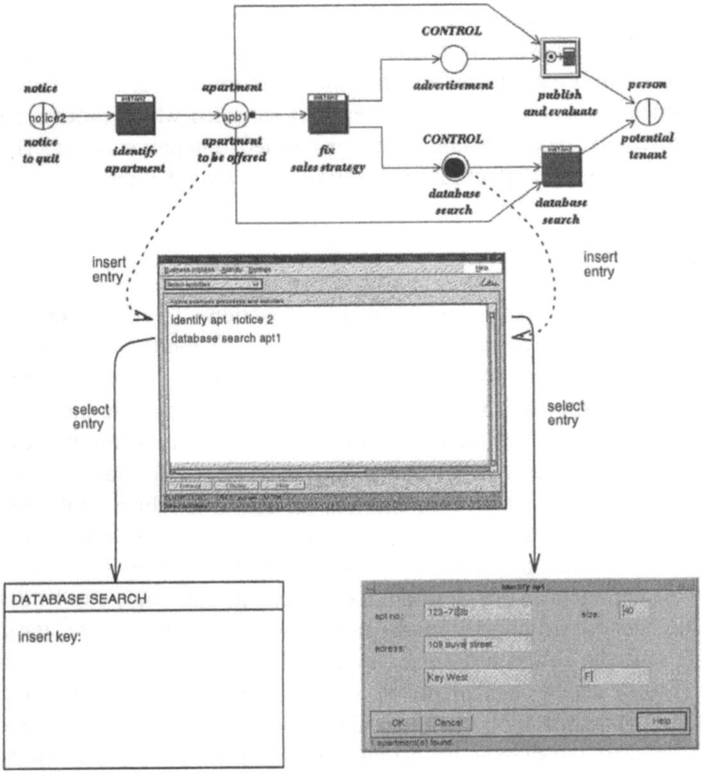

Figure 2. Enaction scheme.

administration industry. Within both projects we introduce further details of the FUNSOFT net approach which were valuable in the projects discussed.

3.1. Process modeling and analysis in software maintenance

The goal of the software process analysis project was to document and to analyze the process of maintaining safety critical software systems. These software systems were simulators of power plants. They were used for training the personnel operating the power plant and for simulating exceptional and accidental situations that could occur when the plant is in operation. Since the system is used as a referential system for the real power plant it is of extreme importance to ensure that the behavior of the simulator is in line with the behavior of the plant. Thus, it is required to update the simulator when the plant is changed or when differences in the behavior of the plant and the simulator are detected.

The software process for maintaining the simulator system was modeled and analyzed in order to

- document the software process as a constructive quality assurance means and to ease the training of those people (and newly hired people) involved in the software processes,
- harmonize the processes for maintaining different simulators such that people could be exchanged between different teams each of which responsible for one simulator,
- try to determine problems in the software processes and to optimize values for size of teams, work packages etc.

The software process to be analyzed was defined after an intensive information analysis (based on specialized questionnaires, interviews and workshops) into which the experts responsible for maintaining the simulator software were involved. This information analysis and the definition of a process model lasted approximately six months. The result was a FUNSOFT net consisting of more than 200 activities, dozens of object types and roles participating in the software process. The net has been discussed and improved with the simulator experts in a couple of reviews.

The process model has been analyzed intensively by simulating it and by evaluating the simulation results. Various simulations have been performed where different parameters such as amount and arrival rate of change requests, amount of and policy for work packages, team size, etc. have been changed. In the following we discuss some example results.

Quantitative evaluations for process planning. During process simulation, simulation traces are recorded (which activities and where executed, how long did they take, which objects and where accessed). Statistical evaluations of simulation traces are used to determine process bottlenecks (channels where the number of tokens exceeds a certain limit), to get information about the degree of parallelism in a process, etc.

In the simulator maintenance process, this statistical evaluation has been used for determining the optimal team sizes, size of work packages, etc. This can be illustrated by a small example: for testing the change request of a work package (each change request has an individual test time request that depends on the complexity of the change request) the simulator software was needed exclusively by the software development department. However, the simulator was also required exclusively by another department which was responsible for training people on the simulator. The access of the two departments to the simulator was regulated by a common schedule. According to that schedule the software development department had certain time slots in which it could access the simulator. Whether or not a change request could be processed, depended on the number of software developers available in the given time slot. Thus, increasing the team size speeded up the process. However, from a certain team size on, the change requests were processed faster than they could be tested. Thus, it could be determined by simulations what the optimal team size was (depending on the parameters such as work package size, activity duration, test time required for change requests, etc.).

Critical path computation. In process modeling, durations of activities can be specified. Figure 3 illustrates the modeled durations on top of the agency symbols. It is possible to define ranges for the duration and it is possible to specify constant durations. During process simulation it is recorded which activity is executed when. In case of a duration range, a value from this range is chosen (following a normal distribution of values

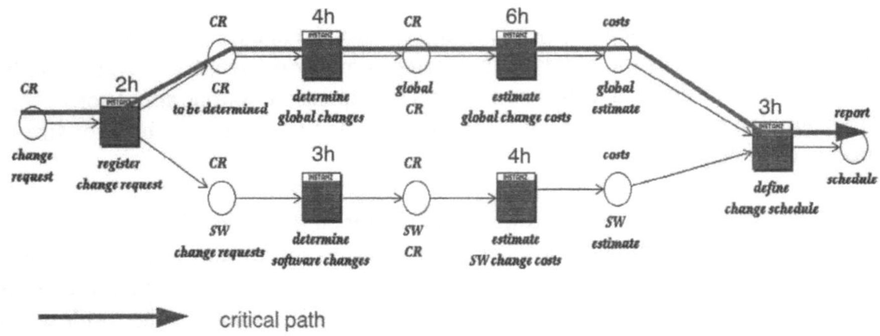

Figure 3. Example of a critical path.

within the range). Starting from the information recorded in a simulation, it is possible to compute the critical path of this simulation, i.e., the sequence of activities that determines the process duration (Deiters and Gruhn, 1991). An example of a critical path is given in figure 3. In that figure an activity model consisting of six activities is shown[4]. Having performed activity *register_change_request* the activity sequences *determine_global_changes* followed by *estimate_global_change_costs* and *determine_software_changes* followed by *estimate_software_change_cost* can be performed in parallel. However, both activities have to be finished for starting activity *define_change_schedule*. Now, considering the activity sequence *determine_global_changes* followed by *estimate_global_change_costs* taking longer than the sequence *determine_software_changes* followed by *estimate_software_change_cost*, the output result from *estimate_software_change_cost* is available prior to the output from *estimate_global_change_costs*. The start of activity *define_change_schedule* is determined by its latest input. That means speeding up the activity sequence *determine_global_changes—estimate_global_change_costs* would speed up the whole process. The critical path of a process is the activity sequence that determines the complete process duration. In figure 3 the critical path is highlighted.

The information about the critical path can be used for indicating:

- where optimizations are reasonable (e.g., by enforcing shorter activity durations) and
- which activities have to be monitored closely during software processes since a delayed termination of those activities would immediately cause a delay in the process.

In the process model for maintaining the simulator the critical path computation has been used for improving coordination between two departments, the software and the hardware department, that both were involved in the software process. In the process the software department was responsible for processing the individual change requests. These change requests were bundled into work packages. Having defined the work packages, it was determined which hardware changes were necessary for implementing the various work packages. However, the hardware department could report that they could not fulfill some of the hardware change requests (for example, because hardware parts could not be provided in the required time frame). This could lead to an iterated rescheduling and replanning of the work packages in the software department. Since the coordination of the two departments on the work packages has been on the critical path, reorganizations of the process have been proposed (e.g., by confirming the possibility to implement the required hardware changes caused by a change request before adopting the change request in the work package).

Using process modeling and analysis techniques, it was possible to document and to improve the maintenance process and by that the planning of the release management of the simulators. However, in order to achieve realistic results all kinds of exceptions had to be adopted in the process model. This led to a substantial complexity of the resulting large-scale process model, especially due to the fact that the processes of several organizational units had to be incorporated into the process model. However, for understanding the process essentials an abstraction from certain process details would have been helpful. It has been shown that there is a tradeoff between modeling all process details for analysis purposes and modeling process abstractions for alleviating a process understanding (see also Section 4.4).

3.2. Process management in housing construction and administration

The starting point for the project *WIS* was the decision of a large German housing construction and administration company to develop a new information system supporting a wide range of business processes related to housing construction and administration (Gruhn and Wolf, 1995). This company had an information system running on a mainframe platform and wanted the new system to be available on a UNIX platform. Once this task was defined, six other companies also bought in. This means, they bought shares of the software system to be developed. All concerned companies formed the WIS consortium. This organization is important because it had impact on the development approach. In more detail, the following aspects had to be considered:

- The requirements analysis was rather difficult, because many different WIS partners had software systems in place already. Generally speaking, nobody wanted to change the business processes he was familiar with.
- WIS partners are rather heterogeneous. This is due to the wide range of numbers of apartments they administrate. A small company, for example, does not use a formalized reminder process, but writes individual reminders, while a large company usually has a detailed and formalized process to handle such cases. Thus, process model customization is necessary for each partner.
- People who had the application domain knowledge (in the area of apartment administration) were not familiar with programming at all. In order to give them the opportunity to participate in the modeling and analysis of business processes to be supported, graphical and intuitive means had to be provided.

Process modeling in this context resulted in large-scale processes. This is due to the many departments of a housing construction and administration company that had to be involved and to the various process ownerships which had to be considered. Figure 4 shows

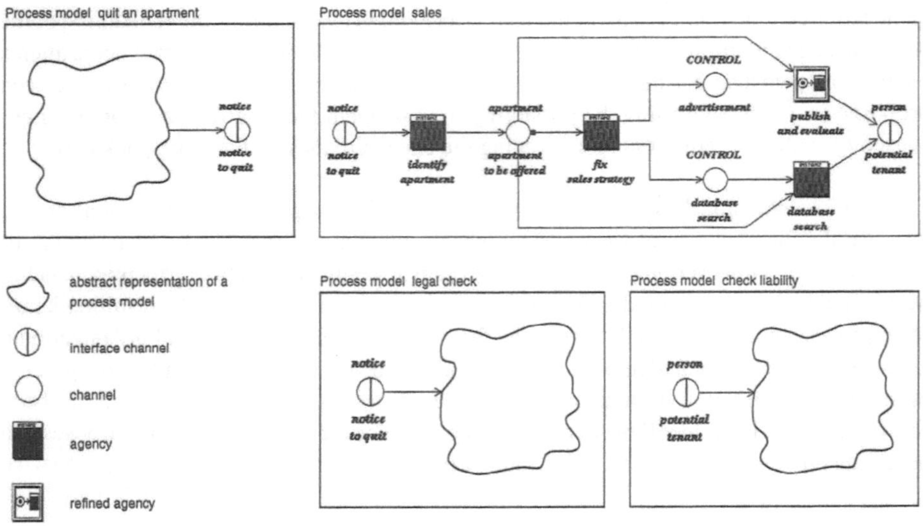

Figure 4. Relationship between process models and their interfaces.

the relationships between processes of a large-scale process as they were identified in WIS. Interfaces between processes are defined by interface channels. An interface channel represents an interface between at least two processes. As soon as an interface channel is filled with an object, this object is forwarded to all other processes which access this channel. Thus, interface channels implement a broadcast mechanism, i.e., a copy of the object is sent to each recipient. There is, for example, an interface channel called *notice to quit*. This interface channel is filled by a process which quits an apartment. The result is an object of type *notice* which is stored in channel *notice to quit*. This channel represents an interface to two other processes.

One of these processes is the sales process, which is notified in order to take all means to rent the apartment again. The first activity to be carried out in this process is the identification of the apartment concerned. Information about this information is written to channel *apartment to be offered*. This channel is accessed by agency *fix sales strategy*. Since the information about the apartment concerned is needed later on (in the activities represented by the agencies *publish and evaluate* and *database search*), agency *fix sales strategy* accesses channel *apartment to be offered* by a COPY-edge (graphically represented by the dot at the edge's foot). This means, that the read object remains in the channel after being read by agency *fix sales strategy*. Within the activity represented by this agency it is decided whether an advertisement for that apartment is to be published or whether the database with potential tenants is to be checked. Depending on this decision, either channel *advertisement* or channel *database search* is filled. At the end of the *sales* process an object of type *person* is written to channel *potential tenant*. Since this channel is an interface channel, the object is forwarded to a *check liability* process.

These relationships between processes concerned by quitting an apartment show that different process participants and different process owners are concerned. Since all these process ownerships have to be considered, we arrive at large-scale processes, which consist of simple processes with clearly defined interfaces.

The processes to be managed in the WIS project are closely tight to the objects being manipulated. A business process for calculating a rent increase has to access the rental agreement, it has to check data describing the rented apartment and possibly a history of earlier rent increases. Due to this close relationship between business processes and their underlying information, the WIS system could be understood as a business process support system based on an information system. To develop this system meant to define the static aspects of the information system (which information is to be stored? how is it related?) and the dynamic aspects (determined by the business processes to be supported). Thus, in addition to the activity-centered perspective of most process management approaches, we also had to consider object-related aspects. In more details this means, that the database, on top of which the business process run, had also to be developed as part of process management.

Another interesting feature of the WIS system is that most business processes in the area of housing construction and administration contain activities in which human actors participate to read, update or delete objects of certain types (e.g., *tenant, apartment, contract*). These activities are usually supported by so-called dialogues. A dialogue consists of a user interface and certain functions linked to the interface. Generally speaking, the processes of WIS are closely related to the underlying information system and they are determined by the manipulation facilities provided by dialogues. Therefore, the FUNSOFT net approach to process management supports the following generation approach:

Database schema generation. The database on top of which the WIS business processes run, has to store objects of all types modeled in the object models. It provides operations to retrieve, update and delete objects of these types. The object type definitions given in the object models contain all information needed to define the database structure and the manipulation operations to insert, update and delete objects of the modeled types. The database generated is based on the database management system Oracle. Roughly speaking, there is one relational table for each object type defined and one relational table for each n : m relationship between object types. More details about the database generation can be found in Gruhn et al. (1995).

Standard dialogue generation. A dialogue to insert an object of type *apartment* is completely determined by the structure of object type *apartment*. Its layout can be generated from the object type definition. Its functionality is defined by the insert, update and delete operations which are generated for each object type modeled. A dialogue generated in this way is called a standard dialogue. In the right bottom corner of figure 5 we recognize a scheme of the layout of a standard dialogue. At the top of a standard dialogue we recognize different menus. They contain operations which have been attached to the object type *apartment*. Moreover, these menus allow one to branch to related objects (according to relationships modeled in the object models) and to father/son objects (according to inheritance relationships modeled in the object models). In the bottom of a standard dialogue we find buttons for inserting new objects and modifying already existing objects. Since some activities require more sophisticated dialogues, which, for example, display objects of different types at one glance, the FUNSOFT net approach supports the editing of individual dialogues.

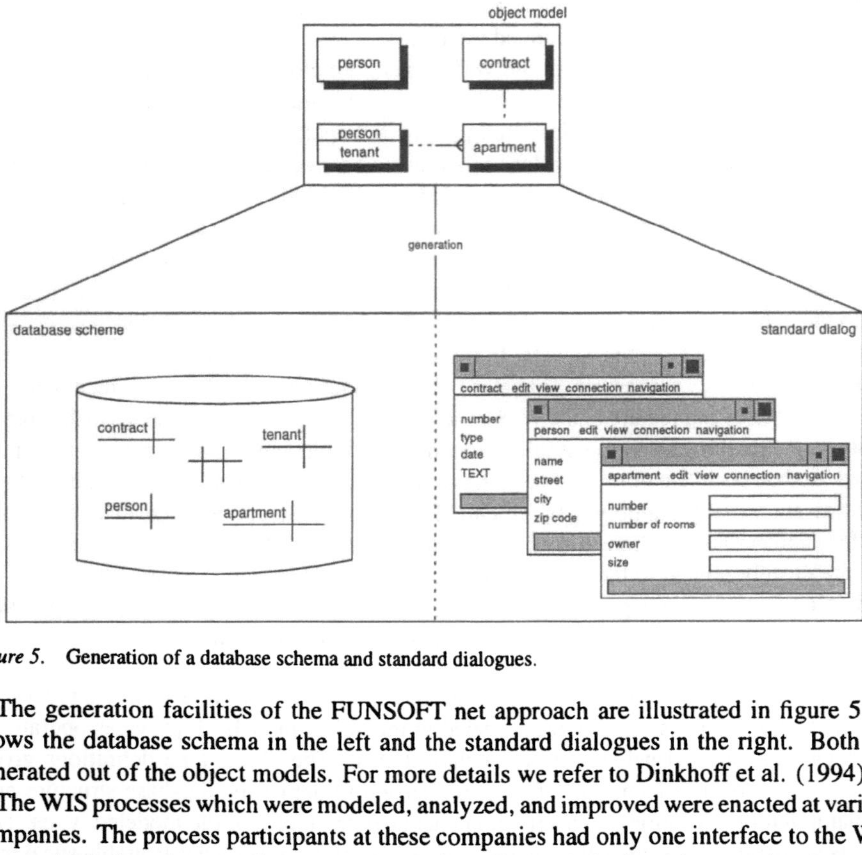

Figure 5. Generation of a database schema and standard dialogues.

The generation facilities of the FUNSOFT net approach are illustrated in figure 5. It shows the database schema in the left and the standard dialogues in the right. Both are generated out of the object models. For more details we refer to Dinkhoff et al. (1994).

The WIS processes which were modeled, analyzed, and improved were enacted at various companies. The process participants at these companies had only one interface to the WIS system. This interface are the personal agendas. Figure 6 displays the agenda of process

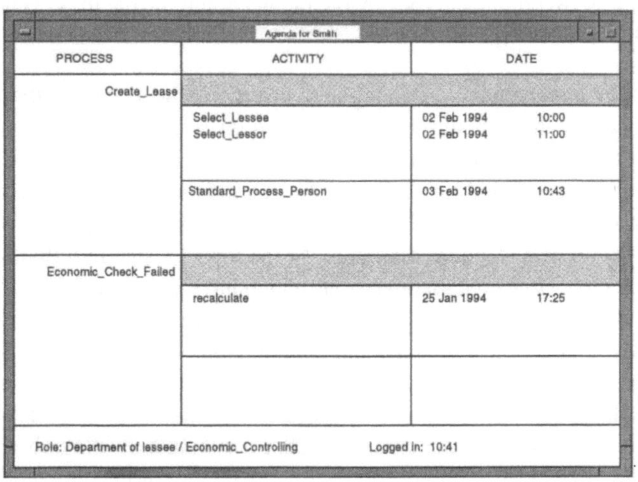

Figure 6. Example of a personal agenda.

participant *Smith*. *Smith* plays certain roles given in the bottom line of the agenda. He is allowed to participate in the processes *Create_Lease* and *Economic_Check_Failed*. The second column shows the activities which could be executed in these processes. The last column shows since when the activities are executable. As soon as *Smith* selects an activity from his agenda, the related dialogue/tool is started at his terminal. Then, the selected activity is deleted from all agendas (because it is under the responsibility of *Smith*. When *Smith* finishes this activity, all activities which become executable are distributed to the agendas of all process participants concerned.

4. Lessons learned and future work

4.1. *Structuring of object models*

At the very beginning of the WIS project a large object model was developed. It turned out that certain processes were related to clearly identified parts of this object model. In order to assign parts of this enterprise-wide object model to process owners, the model was split up into 25 partial object models. Partial object models were glued together by so-called interface object types. An interface object type is defined in one partial object model, exported from that object model and imported into other object models.

4.2. *Maintenance of process model clusters*

Modeling large-scale processes in the form of process model clusters as sketched in figure 4 means to end up with interrelated process models. The modification of one process model can have side-effects on other process models. If, for example, the access to an interface channel is changed, then it can happen, that certain processes are not triggered appropriately any longer. If an interface channel is removed from a certain process model, then it may happen, that certain objects are not processed. In the example discussed, a change in the process model *quit an apartment* could, for example, entail that the sales process is not informed in due time and that, therefore, an apartment is not rented immediately. Due to these interdependencies between process models belonging to large-scale processes we needed several cross-referencing mechanisms whenever modifying interfaces between process models. These cross-referencing mechanisms help to identify which channels are accessed in which process models, which object types are used in which process models, and which dialogues are used for the implementation of which activities.

An additional means to manage modifications of process model clusters is a strict permission concept. By means of this concept we attach the permissions to read and update certain process models and certain interface channels to roles (which themselves can be attached to persons). Thus, it is possible to define that the modeler of the sales process has the permission to model the sales process, but not to modify the interface channel *potential tenant*. Based on these permissions, unambiguous permissions for the maintenance of interfaces between process model can be defined. As soon as an interface is changed all owners of affected process models (i.e., persons who are responsible for the concerned process models) are notified.

4.3. Configuration management for process model clusters

There is a close relationship between process models and mangement of process model configurations. This has been discussed, for example, in (Junkermann et al., 1994; Belkhatir et al., 1993; Gruhn, 1996). In modeling processes from different application domains we came to the conclusion that we should also address the management of configurations of process models. Reasons for doing so are:

Modifications/customization. One crucial requirement for the applicability of a process management approach is that it allows for modifications of process models. Eventhough business processes are not as dynamic as software processes (Ambriola et al., 1990; Bandinelli et al., 1993), they nontheless tend to change over time. Some parts of large-scale processes remain unchanged for their complete lifetime (e.g., nearly all processes that are closely related to legal definitions are more or less fixed and do only change when the legal basis is changed), while others are modified more frequently. The sales process from figure 4, for example, was modified several times. Depending on the general market situation, it was decided not to publish advertisements and to search for potential tenants only by checking the database. Thus, certain parts of large-scale processes had to be modified while others remain unchanged. In the business process environment of WIS it was usually acceptable to wait until no process following the concerned process model was active and to modify the process model at that time. Thus, we did not have to care for sophisticated on-the-fly modification mechanisms, which support the modification process models while processes following these models are active.

Another reason for modifications of parts of large-scale processes in WIS was that process models were developed for a variety of customers. Thus, the process models represented compromises, which had to be customized for some WIS partners. Even if some process models are complete and agreed by all customers, installations at different customers frequently lead to individual adaptations. These adaptations were implemented at the various customer sites. The result were implementations of the WIS system which were slightly different.

Failure diagnosis. It is essential to know which process is running at which customer in order to provide hotline support and remote corrections.

New releases of process models. Only process models which have not been adapted by the customer can be automatically updated by new process model releases. For customer-adapted models, the customer has to choose between keeping his adaptations or getting a new model release (Kaba and Derniame, 1995).

Due to these reasons, configuration management for process model clusters is inevitable. It has to provide support for building configurations of process model clusters from versions of configuration units (e.g., process models, partial data models). It must support versioning of individual process models (instead of considering all process models building one system, e.g., WIS, as one entity to be versioned) and even smaller configuration units. For example, at one customer site one particular dialogue may be changed according to the customer's requirements (without changing the dialogue's input/output behavior). Thus, the structure of the process model remains unchanged and only the concerned dialogue is customer-specific and has to be administrated as a new version. Once that has happened, a new version of the overall process model, which has to be installed at the customer site, has to be merged with the new, customer-specific dialogue version. The same problem applies to the merge of versions of process models which are produced at different sites. As long as the interface of a process model is not changed (i.e., it still reads from and writes to the same interface channels), this model can be combined, with new versions of other process models. Thus customers, can decide for each process model whether or not they want to adapt it without loosing the opportunity for new releases of other process models.

4.4. Transactional modeling

In the process modeling example of the maintenance process for the power plant simulators it has been shown that it is necessary to deal with exceptions in large-scale processes. Among these exceptions are situations that invalidate parts of the work that has been done. The process then has to be continued from a different process state. However, in these situations the process cannot be simply stopped or aborted since activities have been performed whose effects have to be *undone* in order to prevent the process from being in an inconsistent state.

One example from the maintenance process for the power plant simulators is the co-ordination of three different departments. The education department was responsible for training the users on the simulators. They needed the simulators during the training time periods. Thus, deadlines for new releases of the simulators had to be fixed in advance. The software department was responsible for producing the simulator releases. However, since parts of the simulators are hardware parts (lamps, switches, displays etc.), change requests could not only result in software changes, but also in hardware changes. In order to achieve this, hardware change requests have been issued from the software department to a hardware department. Having agreed on a change request, hardware and software department started to work in parallel, the former trying to get or to change the required hardware part and to install it in the simulator then, the latter analyzing what parts of the simulator software had to be changed, designing, implementing and testing the changes afterwards. However, sometimes exceptions could occur where, for example, for processing a change request hardware parts had to be ordered from a supplier that although delivery had been promised could not be delivered in time. Postponing the release was not possible due to the training time period that has been planned on the agreed release date. In this case, it was by all means necessary to *undo* the work that has been done in the software department on that change request for keeping the simulator in a consistent state.

Modeling these kinds of exceptions encompassing the stepping back to earlier process states requires many net elements in the FUNSOFT net approach. This makes the process models complex and, for large-scale process models, the resulting FUNSOFT nets do not have the expressiveness that makes them acceptable as process documentation. To solve this problem, extensions to FUNSOFT nets are needed that allow an easy and flexible mechanism to define how to recover from inconsistent process states. The required modeling constructs and the related mechanisms (aborting process parts, stepping back in a process) are in some sense similar to properties of database transactions. We therefore call the required extensions in our approach FUNSOFT transactions (Deiters et al., 1996).

In the area of workflow management some approaches also adopt transaction concepts of database systems to workflows (Leymann, 1995; Eder and Liebhart, 1996; Weikum, 1993; Wolf, 1994) for ensuring consistency in processes and for supporting an automated error handling in case of failures during error execution. Based on the examination of possible errors, different workflow recovery concepts are developed considering process and document oriented workflows and different kinds of errors like system failures and semantic failures. However, beside managing erroneous situations, the concept of FUNSOFT transaction should also be used as a modeling construct that reduces modeling complexity for certain process situations (e.g., conditions for process state changes in exceptional situations, actions for stepping back in a process).

A description of FUNSOFT transactions can be found in (Deiters et al., 1996). One issue in that concept is the possibility to define that a schedule of activities should run in isolation. That means, all activities that belong to a FUNSOFT transaction are considered to be performed completely or not at all. A comprehensive set of abortion criteria has to be defined in order to indicate when an activity schedule running under transactional properties has to be interrupted and to be reset.

Learned from the example process models described above we consider three types of abortion criteria for a FUNSOFT transaction:

Time constraints. For each transaction a time limit should be defined. If the transaction exceeds the given time limit, the activities of that transaction should not be performed any longer. For example, if in the process model for the maintenance of the power plant simulators the ordering activity of the hardware department would be defined as a transaction, it could be very easily expressed in the process model that after a certain delay in delivery the process needs to be interrupted.

Internal abortion trigger. For each transaction an internal abortion trigger should be defined. This trigger defines an event that can be foreseen in the process model and is caused by one or more activities defined in the model. In the simulator maintenance process model this type of trigger could be used for expressing the need to abort the software changes when the hardware parts are not delivered in time. In this case, the activities of the software department have to be modeled as a transaction. The internal abortion trigger *hardware parts delayed* is raised by the hardware department.

External abortion trigger. It is not always possible to foresee all abortion conditions in advance. In order to deal with unplanned abortion conditions, it should also be possible to define if a transaction can be interrupted or aborted by an authorized person, e.g., the administrator of the process management environment. We call this an external abortion trigger.

The concept of FUNSOFT transactions helps to adopt exceptions in a process model without overloading the model with exception handling. Another issue, of course, is the one of process recovery in strange or erroneous situations. Transactions should allow to specify how to deal with errors that are, for example, made by process participants during the process and how to react in exceptions. On the basis of that specification the process engine can then roll back a transaction when a certain abortion point occurs compensating the effects the transaction has caused so far. In FUNSOFT nets, each agency is annotated with a service defining the functionality (e.g., starting a dedicated tool) to be performed when firing the agency (and executing the corresponding activity). For a more detailed discussion of the enaction architecture we refer to Graw and Gruhn (1995). In order to compensate the effects of the individual activities compensation services are to be specified for each agency. The compensation services are used for rolling back the transaction. Of course, upon roll back of a transaction it cannot be possible to completely undo the effects the various activities of a transaction have been performed. In these cases the effect of a compensation service must rather be the notification that a produced result (e.g., the print out of a document) is recognized as invalid.

5. Conclusion

The FUNSOFT net approach had been applied to academic projects. This, already, had resulted in improvements of the approach. After several experiments the approach seemed to be consolidated. Nontheless, industrial experiments (which have been started after this first consolidation phase) revealed further requirements. These requirements were related to size and complexity of process models, number of process modelers to be supported, and integration of process models with software systems already in use. Dealing with these requirements resulted in a second consolidation phase.

This experience underpins the idea that academic experiments are sufficient to prepare technology transfer, but that they have to be complemented by industrial projects. Related to the FUNSOFT net approach, academic and industrial projects yielded insights which finally helped to implement the FUNSOFT net approach in a commercial process management environment, called LEU (Dinkhoff et al., 1994).

By now, the FUNSOFT net approach has been applied to highly structured processes (i.e., it is precisely defined what has to happen when and by whom). Other, more unstructured types of processes demand for less prescriptive definitions and for less strict process enaction. In the future, we will check which extensions of the FUNSOFT net approach are necessary to also manage unstructured processes.

Notes

1. The term *process* is used in the following as a generalization of business process and software process. Whenever we refer to a particular specialization we use the term *business process* or the term *software process*.
2. Figure 1 shows only a scheme of an integrated process model. This scheme does not reveal the detailed syntax of any of the languages used, but it illustrates the basic principle of integration of organization models, activity models and object models.
3. An agenda example is discussed in detail in Section 3.2, figure 6.
4. Since the example is used for illustrating a critical path and not for describing details for the simulator maintenance process we do not discuss the activities more thoroughly here.

References

Adomeit, R., Deiters, W., Holtkamp, B., Rockwell, R., Schülke, F., and Weber, H. 1994. *Software Engineering Environments.*

Ambriola, V., Ciancarini, P., and Montangero, C. 1990. Software process enactment in Oikos. *Proceedings of the Fourth ACM SIGSOFT Symposium on Software Development Environments*, Irvine, California, USA, pp. 183–192. Appeared as ACM Software Engineering Notes, 15(6).

Bandinelli, S., Fuggetta, A., and Ghezzi, C. 1993. Software process model evolution in the SPADE Environment. *IEEE Transactions on Software Engineering*, 19(12):1128–1144.

Bandinelli, S., Braga, M., Fuggetta, A., and Lavazza, L. 1994. The architecture of SPADE-1 process-centered SEE. In *Software Process Technology—Proceedings of the 3rd European Software Process Modeling Workshop*, B. Warboys (Ed.), Villard de Lans, France: Springer. Appeared as Lecture Notes in Computer Science 772.

Belkhatir, N., Estublier, J., and Melo, W.L. 1993. Software process model and work space control in the Adele system. *Second International Conference on the Software Process*, In L. O'Conner (Ed.), Berlin, Germany, IEEE Computer.

Broeckers, A. and Gruhn, V. 1993. Computer-aided verification of software process model properties. *Proceedings of the Fifth Conference on Advanced Information Systems Engineering (CAiSE)*, Paris, France.

Bunje, T.D., Engels, G., Groenwegen, L., Matsinger, A., and Rijnbeek, M. 1996. Industrial maintenance modelled in SOCCA: An experience report. *Proceedings of the 4th International Conference on the Software Process*, Brighton, UK.

Deiters, W. 1993. A view based approach to software process management. Ph.D. thesis, University Dortmund. Appeared as Technical Report No. 467/93.

Deiters, W. and Gruhn, V. 1991. Software process model analysis based on FUNSOFT nets. *Mathematical Modeling and Simulation*, (8).

Deiters, W. and Gruhn, V. 1994. The FUNSOFT net approach to software process management. *International Journal of Software Engineering and Knowledge Engineering*, 4(2):229–256.

Deiters, W., Lindert, F., and Schiprowski, R. 1996. A transaction concept for FUNSOFT nets. In *Proceedings of the Workshop Business Process Modelling and Workflow Management, Annual Conference of the German Computer Society's (GI) 1996-Klagenfurt*, R. Oldenbourg (Ed.), Vienna, Austria.

Derniame, J.-C. and Gruhn, V. 1994. Development of process-centered IPSEs in the ALF project. *Journal of Systems Integration*, 4(2):127–150.

Dinkhoff, G., Gruhn, V., Saalmann, A., and Zielonka, M. 1994. Business process modeling in the workflow management environment LEU. In *Proceedings of the 13th International Conference on the Entity-Relationship Approach*, P. Loucopoulos (Ed.), Manchester, UK: Springer, pp. 46–63. Appeared as Lecture Notes in Computer Science no. 881.

Eder, J. and Liebhart, W. 1996. Workflow recovery. *First IFCIS International Conference on Cooperative Information Systems (CoopIS96)*, IEEE Computer Society Press: Brussels, Belgium.

Emmerich, W., Bandinelli, S., and Lavazza, L. 1996. Fine-grained process modelling: An experiment at British Airways. *Proceeding of the 4th International Conference on the Software Process*. Brighton, UK (to appear at ICSP4).

Engels, G. and Groenewegen, L. 1994. Specification of coordinated behavior by SOCCA. In *Software Process Technology—Proceedings of the 3rd European Software Process Modeling Workshop*, B. Warboys (Ed.), Villard de Lans, France: Springer, pp. 128–151. Appeared as Lecture Notes in Computer Science 772.

Finkelstein, A., Kramer, J., and Nuseibeh, B. (Eds.) 1994. *Software Process Modelling and Technology*, New York, US: John Wiley and Sons.

Genrich, H.J. 1987. Predicate/transition nets. In *Petri Nets: Central Models and Their Properties*, W. Brauer, W. Reisig, and G. Rozenberg (Eds.), Berlin, FRG: Springer, pp. 208–247. Appeared in Lecture Notes on Computer Science 254.

Ghezzi, C., Jazayeri, M., and Mandrioli, D. 1991. *Fundamentals of Software Engineering*. Englewood Cliffs, New Jersey: Prentice-Hall.

Graw, G. and Gruhn, V. 1995. Distributed modelling and distributed enaction of business processes. *Software Engineering—ESEC'95, 5th European Software Engineering Conference*, Sitges, Spain: Springer, pp. 8–27. Appeared as Lecture Notes in Computer Science no. 989.

Gruhn, V. 1991. Validation and verification of software process models. Ph.D. thesis, University of Dortmund. Appeared as Technical Report No. 394/91.

Gruhn, V. 1996. Configuration management support for software process models. In *Software Process Technology—Proceeding of the 5th European Workshop, EWSPT96*, C. Montangero (Ed.), Nancy, France: Springer, pp. 132–136. Appeared as Lecture Notes in Computer Science 1149.

Gruhn, V. and Wolf, S. 1995. Software process improvement by business process orientation. *Software Process Improvement and Practice*, 1:49–56 (Pilot Issue).

Gruhn, V., Pahl, C., and Wever, M. 1995. Data model evolution as a basis of business process management. In *OOER95: Object-Oriented and Entity-Relationship Modeling*, Gold Coast, Australia: Springer Berlin, pp. 270–281. Appeared as Lecture Notes in Computer Science 1021.

Hales, K. and Lavery, M. 1996. *Workflow Management Software: The Business Opportunity*. London, UK: Ovum Ltd.

Hammer, M. and Champy, J. 1993. *Reengineering the Corporation*. New York, US: Harper Business.

Junkermann, G., Peuschel, B., Schäfer, W., and Wolf, S. 1994. MERLIN: Supporting cooperation in software development through a knowledge-based environment. In *Software Process Modelling and Technology*, B. Nuseibeh, A. Finkelstein, and J. Kramer (Eds.), Somerset, England: John Wiley and Sons, pp. 103–129.

Kaba, A.B. and Derniame, J.-C. 1995. Transients change processes in process-centered environment. In *Software Process Technology—Proceedings of the the 4th European Software Process Modeling Workshop*, W. Schäfer (Ed.), Noordwijkerhout, Netherlands: Springer, pp. 255–259. Appeared as Lecture Notes in Computer Science 913.

Leymann, F. 1995. Supporting business transactions via partial backward recovery in workflow management systems. In *GI Fachtagung: Daten-Banksysteme in Büro, Technik und Wissenschaft*, G. Lausen (Ed.), Dresden, Germany: Springer, pp. 51–70. Appeared as Informatik Aktuell.

Lonchamp, J. 1994. An assessment exercise. In *Software Process Modelling and Technology*, B. Nuseibeh, A. Finkelstein, and J. Kramer (Eds.), Somerset, England: John Wiley and Sons, pp. 335–356.

Schäfer, W. (Ed.) 1995. *Software Process Technology—Proceedings of the 4th European Workshop on Software Process Modelling*, Noordwijkerhout, The Netherlands: Springer. Appeared as Lecture Notes in Computer Science 913.

Totland, T. and Conradi, R. 1995. A survey and comparison of some research area relevant to software process modelling. In *Software Process Technology—Proceedings of the 4th European Software Process Modeling Workshop*, W. Schäfer (Ed.), Noordwijkerhout, Netherlands: Springer, pp. 65–69. Appeared as Lecture Notes in Computer Science 913.

Warboys, B. 1994. Reflections on the relationship between BPR and software process modelling. In *Proceedings of the 13th International Conference on the Entity-Relationship Approach*, P. Loucopoulos (Ed.), Manchester, UK: Springer, pp. 1–9. Appeared as Lecture Notes in Computer Science no. 881.

Weikum, G. 1993. Extending transaction management to capture more consistency with better performance. In *Proceeding of the 9th French Database Conference*, Toulouse, France.

Wolf, S. 1994. Ein transaktionsbasierter Ansatz zur Unterstützung kooperrativer Softwareentwicklung. Ph.D. thesis, University of Dortmund.

Automated Software Engineering 5, 27–60 (1998)

Serendipity: Integrated Environment Support for Process Modelling, Enactment and Work Coordination

JOHN C. GRUNDY jgrundy@cs.waikato.ac.nz
Department of Computer Science, University of Waikato, Private Bag 3105, Hamilton, New Zealand

JOHN G. HOSKING john@cs.auckland.ac.nz
Department of Computer Science, University of Auckland, Private Bag 92019, Auckland, New Zealand

Abstract. Large cooperative work systems require work coordination, context awareness and process modelling and enactment mechanisms to be effective. Support for process modelling and work coordination in such systems also needs to support informal aspects of work which are difficult to codify. Computer-Supported Cooperative Work (CSCW) facilities, such as inter-person communication and collaborative editing, also need to be well-integrated into both process-modelling tools and tools used to perform work. Serendipity is an environment which provides high-level, visual process modelling and event-handling languages, and diverse CSCW capabilities, and which can be integrated with a range of tools to coordinate cooperative work. This paper describes Serendipity's visual languages, support environment, architecture, and implementation, together with experience using the environment and integrating it with other environments.

Keywords: process modelling, process enactment, process-centered environments, work coordination, environment integration

1. Introduction

Most computerised or semi-computerised work systems have evolved informal or semi-formal process models. These attempt to describe the use of different tools on a project, the interchange and modification of work artefacts by tools and workers, and the flow of control and/or data. Workflow Management Systems (WFMS) and Process-Centred Environments (PCEs) are examples of tools developed to assist in the construction, use and evolution of formalised process models. While many of these systems support modelling and enacting work processes well, they have some deficiencies. Some use low-level, textual process models which are difficult for end-users to understand and modify (Swenson, 1993; Bogia and Kaplan, 1995), some lack support for inter-person communication (Di Nitto and Fuggetta, 1995; Ben-Shaul and Kaiser, 1996; Tolone et al., 1995), and many provide either inflexible process models, which inadequately handle exceptions and are difficult to change while in use (Swenson et al., 1994; Tolone et al., 1995), or do not adequately support the more informal aspects of cooperative work (Kaplan et al., 1996; Tolone et al., 1995).

Computer-Supported Cooperative Work (CSCW) systems provide communication and cooperative editing tools to support collaborative work. These include synchronous editors and tools, such as shared workspaces, telepointers and video conferencing, and asynchronous communication, such as e-mail and version control systems (Ellis et al., 1991). However, many of these systems focus on low-level collaborative editing and/or person-to-person communication issues, and lack support for general work process modelling (Krishnamurthy and Hill, 1995). Some systems, such as ConversationBuilder (Kaplan et al., 1992a), SPADE/ImagineDesk (Di Nitto and Fuggetta, 1995), and Oz (Ben-Shaul and Kaiser, 1994a), attempt to bridge the gap between CSCW tools and process modelling tools but with limited success. Others, such as wOrlds (Bogia and Kaplan, 1995), provide facilities for supporting the informal aspects of work but do not adequately support codified, cooperative process models (Kaplan et al., 1996). Most existing CSCW, WFMS and PCEs tend not to be well integrated with each other, nor with work tools, such as document editors, CASE tools and programming environments (Tolone et al., 1995; Ben-Shaul and Kaiser, 1996). This usually means custom-built systems must be developed which exhibit CSCW behaviour, or existing tools modified to permit (often loose) integration with process modelling tools.

We describe Serendipity, which integrates process modelling and cooperative work support for large collaborative systems. Serendipity provides a novel graphical process modelling language which can be used to represent both general process models and plans for histories of work. Another graphical language specifies event handling for process model enactment and work artefact modification events. We focus on the design of these languages and a supporting environment for them, together with our experiences in integrating this environment with other systems, including both CSCW tools and tools for performing work, such as CASE, programming environment, and office automation tools. Serendipity provides a "work context" for these tools, and supports high-level work context awareness and work coordination for them. Descriptions of work artefact changes from integrated tools are annotated with work context information and stored by the current enacted process stage, along with process enactments, forming work and enactment histories. Serendipity is integrated with existing tools via an event passing system, necessitating no modifications to these tools. Serendipity has been used for collaborative process modelling and work planning, process improvement, method engineering and office automation.

2. Problem domain: Large cooperative work systems

Tools that effectively support collaborative work in large problem domains, where several people, tools and many work artefacts are involved, such as software development or office automation, have a number of important requirements:

- Support for modelling work processes (Barghouti, 1992; Swenson et al., 1994; Tolone et al., 1995). This allows cooperating people to more readily plan and coordinate their work on large problems. Approaches to providing such process modelling range from precise, formal languages (Barghouti, 1992; Ben-Shaul and Kaiser, 1994a; Bandinelli, et al., 1994) to more high-level, graphical workflow languages (Medina-Mora et al., 1992; Swenson et al., 1994; Baldi et al., 1994). Process models should ideally allow work processes to be enforced or used as guidance, and should be readily understandable and modifiable by users.

- The ability to handle arbitrary events relating to process model state changes, work artefact updates or events in work tools (Bogia and Kaplan, 1995; Grundy et al., 1995a). When the state of a process model or work artefact changes, interested collaborators may need to be informed, or automatic responses carried out (Ben-Shaul and Kaiser, 1994a; Bogia and Kaplan, 1995).
- Effective inter-person communication and collaborative editing support is needed (Di Nitto and Fuggetta, 1995; Ben-Shaul and Kaiser, 1996). With process models providing a high level of "work context" information, keeping collaborators aware of each others' work is necessary (Bogia and Kaplan, 1995). Allowing people to intermittently join or leave cooperative work sessions, or to review the reasons for artefact changes, requires annotated histories of work for process stages.
- Tools for performing work need to be well-integrated with the tools for communicating and editing work artefacts (CSCW tools) and the process modelling tools (WFMS and PCEs) (Di Nitto and Fuggetta, 1995; Ben-Shaul and Kaiser, 1996; Kaplan et al., 1996). This should allow: CSCW tools to be utilised to discuss and annotate both work and process model artefacts; the process modelling tools to control CSCW and work tool behaviour; work performed with the work tools to be readily coordinated by the process modelling tools.

As an example of a process modelling language and environment, consider Swenson's Visual Planning Language (Swenson, 1993), and its embodiment in support environments Regatta (Swenson et al., 1994) and TeamFLOW (TeamWARE, 1996). Figure 1 shows a VPL model describing a process to modify a software system. Ovals are process stages, hexagons start/stop states, and flows between stages represent finishing states of one stage which enact the linked stages.

VPL has many advantages over comparable process modelling languages in that it is concise and versatile. It can be used to both model processes and plan work on specific projects, and Regatta and TeamFLOW allow their process models to be enacted by multiple collaborating users. However, VPL does not adequately model work artefacts or tools, using only simple textual attributes of process stages to indicate this information. It does not support the handling or arbitrary events from related process stages, nor tools and artefact updates. The integration of its supporting environments with tools for performing work is limited. For example, TeamFLOW is unable to handle events from work tools or directly control their usage, other than invoking them, when appropriate, from process model stages.

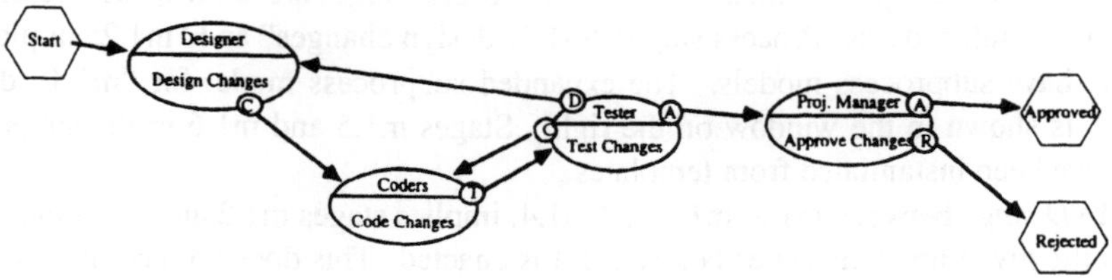

Figure 1. An example VPL process model.

Some work has been done on integrating PCE/WFMS, CSCW and work tools. Examples include ConversationBuilder (Kaplan et al., 1992a), wOrlds (Tolone et al., 1995), SPADE and ImagineDesk (Di Nitto and Fuggetta, 1995), Marvel and ProcessWEAVER (Heineman and Kaiser, 1995), and Oz (Ben-Shaul et al., 1994b; Ben-Shaul and Kaiser, 1994a, 1996). These approaches have produced useful environments with some integration of PCEs/WFMSs, CSCW tools and tools used to perform work. However, this work has so far not produced the "ideal" integration of these technologies, due to problems with adequately integrating disparate tools (Tolone et al., 1995; Valetto and Kaiser, 1995; Di Nitto and Fuggetta, 1995), or with limited scope of the software development and/or workflow tools used (Kaplan et al., 1996; Di Nitto and Fuggetta, 1995).

Serendipity is our approach to solving these problems. It extends VPL to more completely describe work processes, and adds a new visual language for specifying event handling. The next two sections describe Serendipity's visual languages, and are followed by a description of the support environment. Section 6 describes integration of Serendipity with other tools, and is followed by a description of the architecture and implementation. Section 8 describes experience using Serendipity, compares it with other languages and systems, and discusses current and future work.

3. High-level process modelling and work planning

3.1. Extended visual planning language

We have developed the Extended Visual Planning Language (EVPL), which preserves the notion of simple "work plans" from VPL, but adds capabilities to support general process modelling. EVPL can thus be used to both model generic, reusable work processes, and to model and record work plans and histories for a particular project. New notational components in EVPL include: identifiers for process stages, which allow stages to be more readily identified and referenced by other process models; role, artefact and tool representations, so these aspects of the work context for a process stage can be specified; "usage" connections between process stages and role, artefact and tool representations; and various annotations on usage connections to indicate how these other parts of a work context are utilised by process stages. Window "m1: model1-process" in figure 2 uses EVPL to describe a simple software process model for updating a software system. Figure 3 summarises the basic EVPL modelling capabilities.

The figure 2 example would normally be part of a larger process, such as the ISPW-6 software process example (Kellner et al., 1990). Several stages are shown, each describing part of the overall process. Process stages "m1.1: design changes" and "m1.2: implement changes" have subprocess models. The expanded subprocess model for "m1.1: design changes" is shown in the window on the right. Stages m1.5 and m1.6 in the subprocess model have been instantiated from templates.

The *AND stage* between m1.2, m1.3 and m1.4, implies stages m1.2 and m1.3 must both finish in the given finishing states before m1.4 is enacted. This does not necessarily mean all subprocesses of m1.2 or m1.3 need be finished, as they may have parallel flows which terminate with different finishing states at differing times. However, at least one flow must

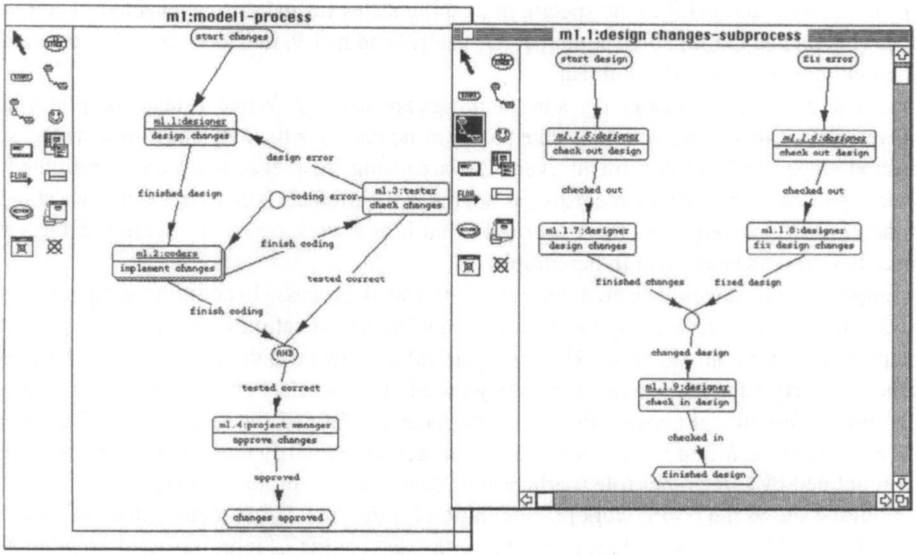

Figure 2. A simple software process model and subprocess model in Serendipity.

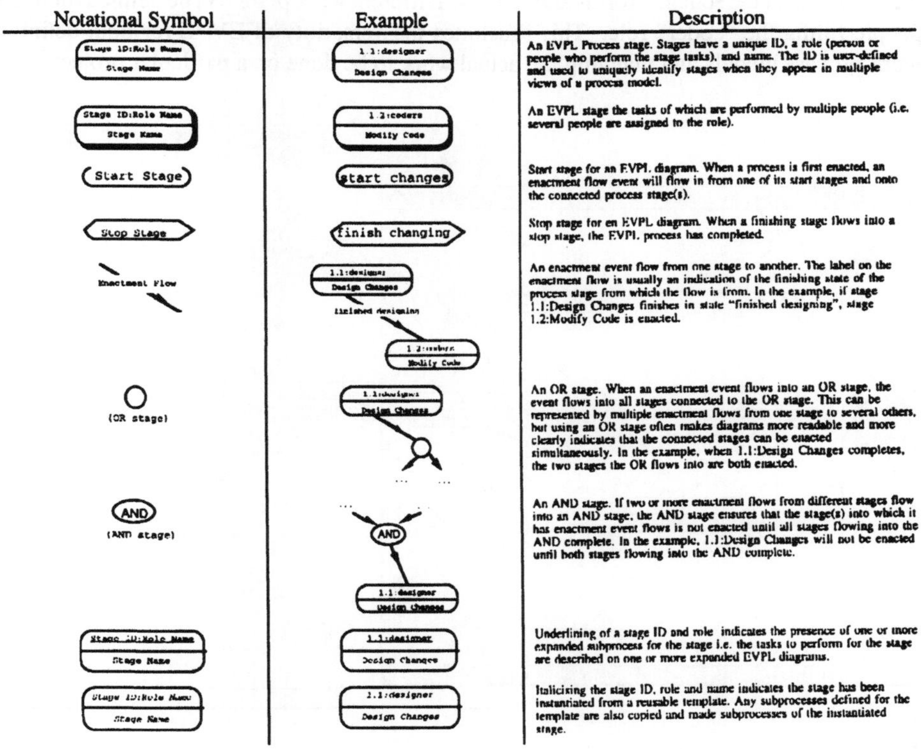

Figure 3. Basic modelling capabilities of EVPL.

terminate m1.2 and m1.3 in the specified finishing states for m1.4 to be enacted. The *OR stage* (unlabelled circle) connecting m1.1.7, m1.1.8 and m.1.9, indicates m1.1.9 is enacted when either m1.1.7 or m1.1.8 finish.

To use this process model, stages in the model are *enacted.* When a stage completes in a given state, event flows with that state name (or no name) activate to enact linked stages; enacted stages and the enactment event flows causing their enactment are highlighted. Enactments of each stage are recorded, as are the work artefact changes made while a stage is the *current enacted stage* for a user (i.e., that user's work context). Section 5 further describes process enactment in Serendipity.

Empty, or leaf, stages have no definition of their work process. In addition, some process model data, such as roles, may be abstract, requiring specification for a particular project the process is used on. Users working on a particular project can define work plans for leaf stages, and extend models to more precisely specify role, artefact and tool data. For example, the process model above indicates work plans are defined for each "coder" for stage m1.2. The people filling these "coder" roles can define or modify these work plans or have them defined by some other role (perhaps the "designer" or "project manager").

Figure 4 shows the EVPL work plan definition for the "m1.2: implement changes" stage for coder "john". When this stage is enacted with "start coding", "john" is to add a relational database table "address", modify a table "customer", and then modify any forms and reports affected by these schema changes. If the stage is enacted with "fix code", "john" must locate the error and fix the schema, forms and reports. Different work plans can be defined for other people filling the "coder" role. This illustrates the versatility of EVPL for both defining work process models and planning the actual work to be done on a particular project.

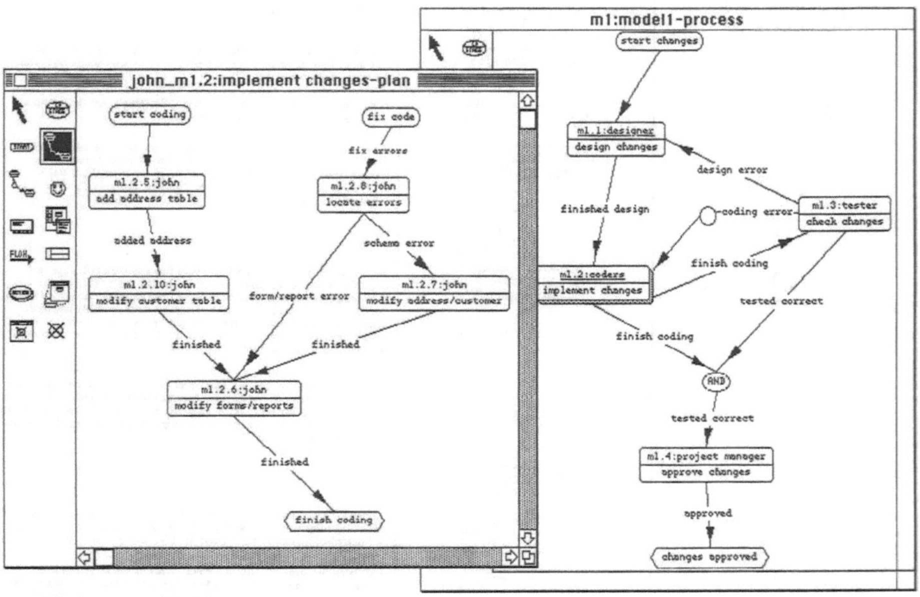

Figure 4. A work plan for coder "john".

Note that there is a mapping from the "finished design" and "coding error" finishing states of m1.1 and m.1.3 to the "start coding" and "fix code" starting state names of the subprocess model. This mapping is specified in a dialog when adding each enactment event flow. Stages can be enacted in the same start state by multiple event flows, such as stage m1.2.6 in figure 4.

3.2. *Describing process stage work context information*

EVPL also extends VPL to permit capture of *work context* information in the form of tools and artefact used in each stage, and coordination and communication needed between stage roles (figure 5). Some, but not all, of these additional modelling capabilities are found in process modelling languages developed by other researchers (for example, E3 PML (Baldi et al., 1994)).

Figure 6 shows a different perspective (view) of the process model of figure 2. This retains the process stages, but not the enactment event flows. Instead, usage flows describe the tools, artefacts and roles that the stages use. For example, "m1.1: design changes" uses the "ER Modeller" tool and the "ER design" artefact. The usage flow between "ER design" and "ER Modeller" specifies the latter is used to modify the former. The usage connections to the "designer" role from m1.1 and m1.3 indicate coordination between the m1.3 role ("tester") and the m1.1 role ("designer"), in this case the roles must communicate informally ("talks with"). Annotations on the usage flows indicate, for example, that the designer must use the ER modeller to design changes (√), the ER design artefact is updated by "m1.1: design changes" (U), the changes list artefact is created and/or updated by "m1.1: design changes" (CU), m1.2: implement changes accesses (but does not update) the ER design (A), and process stages m1.2 and m1.3 cannot be enacted at the same time as process stage m1.4 (¬).

Notational Symbol	Example	Description
artefact	class	These describe the kinds of work artefacts used by process stages. Artefacts can represent a general class of work artefact (e.g. "classes" in an OO system), or specific instances of artefacts (e.g. "class 'window'").
ToolName	ER Modeller	Tools are used to modify or view work artefacts. If a meta-process model is being defined, tools may include the Serendipity environment itself, and artefacts may include process model components.
RoleName	john	Roles represent people or abstract roles associated with process stages, tools or artefacts. Roles may refer to specific people (e.g. "john"), groups of people ("coders"), or abstract roles (e.g. "designer")
UsageFlowLabel	m1.1:designer design changes; used by; ER Modeller	Usage connections indicate that an artefact, tool, role or process stage is used by another. The flow is from the used process model component to the using process model component.
(CAUD√¬)	m1.1:designer design changes; creates (CU); changes list	Various annotations can be added to a usage flow: 'C' = using process model component creates instances of used component; 'A' = using component accesses used component; 'U' = using component updates used component; and 'D' = using component deletes instances of used component. C, U and D are generally only useful when the used component is an artefact. √ specifies the using component must use the used component, whereas ¬ specifies the using component must not use the used component. For process stages, ¬ specifies non-concurrent enactment, √ concurrent enactment.

Figure 5. Extra modelling capabilities of EVPL to describe work contexts.

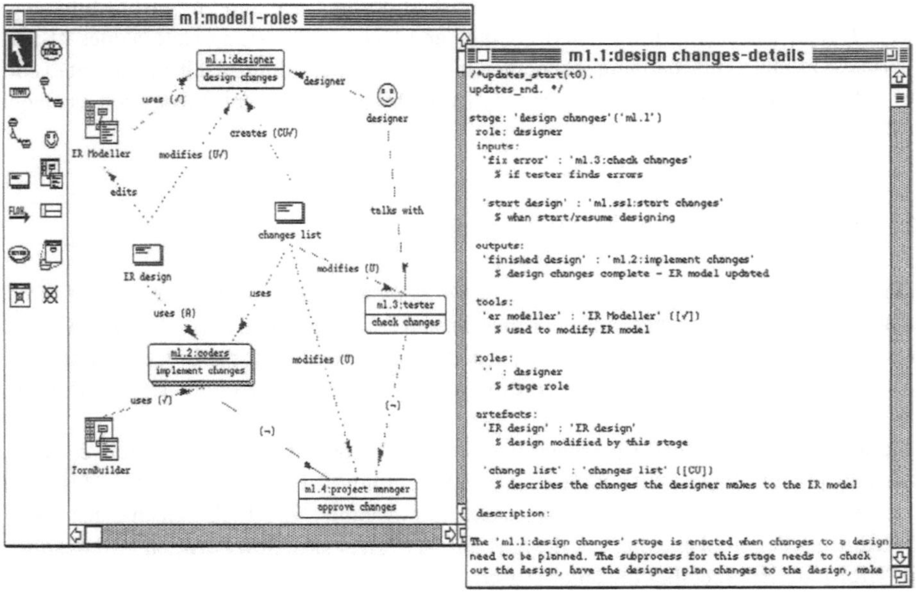

Figure 6. A data, tool and role-oriented perspective of the first process model (left) and a textual stage view (right).

3.3. *Textual process information*

The graphical process model, work plan and work context views described above, can be supplemented using textual representations of process stage, artefact, tool and role information. An example is given in figure 6. Extra information which can be specified includes: user-defined attributes for any kind of process model component (for example, a "percent_complete" value for a process stage); user-defined comments about process model component information; and bindings of abstract role to concrete role (e.g., that "john" will fill the "designer" role).

4. Defining event-handling filters and actions

The process models described in the previous section are fairly static. Stage enactment flows are described, which indicate the flow of enactment events between stages. Often, however, other types of event handling are required, for example to specify: dependencies between process stages which belong to different process model views; the handling of artefact update and/or tool events; automatically applied rules or constraints on process models, driven by event triggers; and the automatic invocation of inter-person communication tools.

Few graphical process modelling languages support such capabilities. Most typically use some form of textual specification of process model stage attributes to specify limited forms of event handling, or utilise complex rule-based languages. Neither of these approaches capture and represent graphically the kinds of events nor how they are handled. To this end we developed the Visual Event Processing Language (VEPL), to permit visual specification of arbitrary event handling and event-triggered process model rules.

4.1. *Basic event-handling*

The basic VEPL constructs are *filters* and *actions*, which receive *events* from stages, artefacts, tools or roles, or other filters and actions. Filters match received events against user-specified criteria, passing them onto connected filters and actions if the match succeeds. When actions receive an event, they carry out one or more operations in response to the event. These operations update information and/or generate new events, which may be detected and acted upon by other filters and actions. VEPL is fully integrated with EVPL and thus provides a graphical, high-level specification of event handling for EVPL process models and work plans.

Figure 7 shows the main components of VEPL. Some of these, such as enactment event flows and usage connections, are derived from EVPL, but have different semantics when used with VEPL filters and actions.

Figure 8 shows a simple event-handling view which extends the process model from figure 2, illustrating use of VEPL for inter-stage work coordination. In this example, testing of software (stage m1.3) can be carried out while further design and/or coding takes place. To coordinate the people associated with the "designer", "coders" and "tester" roles, figure 8 specifies when notifications are sent to coders that testing has been completed or is in progress.

The lefthand event flow from m1.3 specifies that all enactment events on m1.3 should flow into the filter "Made Current". This checks if the received event shows the stage has been made the user's current enacted stage (i.e., the user is now working on this stage) and, if so, the event flows into the OR stage and on to "Not Completed"; otherwise event propagation

Notational Symbol	Example	Description
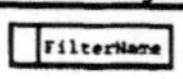	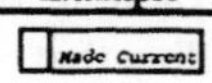	A filter definition. Filters receive events (from process stages, artefacts, tools, roles, other filters or actions) and if the event matches the defined selection criteria for the filter, the event is passed onto the connected filters and/or actions. A filter or action reused from a template filter/action definition has its name in italics
		An action definition. Actions receive events (from process stages, artefacts, tools, roles, filters or other actions) and respond to the event by performing some action (which often generates other events). Actions can pass on events to other filters and/or actions.
	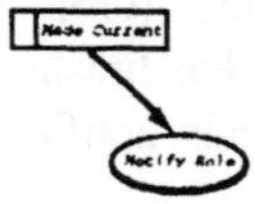	An event flow into/from a filter or action. Events may be process stage enactment events, artefact update events, tool events, some event caused by a role (i.e . user), or an event generated by an action. For example, if Made Current decides an enactment event flowing into it means a process stage has been made the current enacted stage, then the Notify Role action is invoked to notify another user about this event.
		Usage flow into a filter or action. These specify parameters of the filter/action. For example, the MadeBy filter is parameterised by a role name which it uses to decide whether some event was caused by a particular role. In the example, that role name is instantiated to "john" by the usage connection to the role process model component.

Figure 7. Basic modelling capabilities of VEPL.

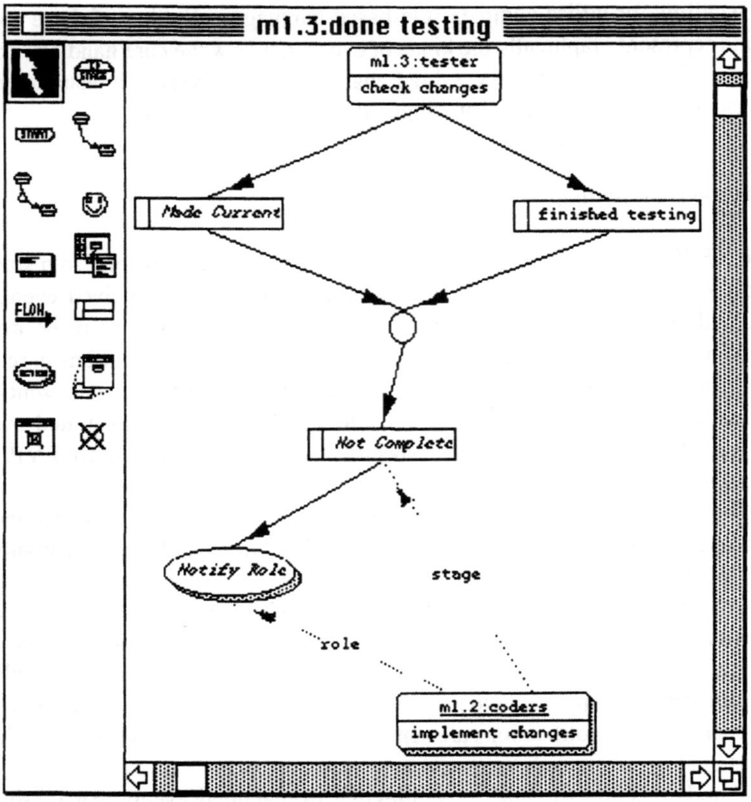

Figure 8. An example of process stage event filtering.

stops. "Not Completed" is parameterised by a stage id, instantiated in this case to stage m1.2. This filter checks its stage parameter is either enacted or able to be restarted (not yet completed) and, if so, passes received event flows into action "Notify Role". This action, parameterised by a role, informs the person or people filling that role (in this example, the "coders" of m1.2), that testing has started or finished. Notification is, by default, via an e-mail like message, but users can specify, via dialogue box, they wish to be informed in other ways, such as by opening a dialog on the coder's' screen, shading the "m1.2: implement changes" stage icon to indicate presence of a message, etc. (Grundy et al., 1996a). The righthand event flow from m1.3 notifies coders if testing has finished in the "fix code", "fix design" or "finished testing" states. The coders are again notified only if the m1.2 stage has not completed.

"Notify Role", "Made Current" and "Not Complete" are reusable (system defined) filters/actions while "finished testing" is specific to this particular process model. Filters are defined by a pattern-matching language (for simple filters), instantiation of a reusable template (via a forms interface), or use of an API to Sernipity's implementation language (see later in this section).

4.2. *Handling artefact, hierarchical and multiple events*

To allow different processing based on the kinds or numbers of events, VEPL allows users to: distinguish between enactment and tool/artefact update events; specify if subprocess component events are handled; and specify that a filter/action is to handle sequences of, rather than individual, events. Figure 9 shows these additional modelling capabilities.

As an example, figure 10 specifies that the "designer" associated with "m1.1: design changes" is to be notified of any schema updates made by "coders" associated with "m1.2: implement changes". Artefact updates (indicated by the Æ) made by any user filling the "coder" role while working on m1.2, or any of its subprocesses (indicated by the ∑), flow into a filter ("RDB table change"), which checks if the change is to a table. If so, the

Notational Symbol	Example	Description
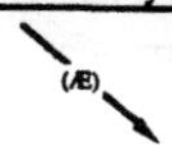	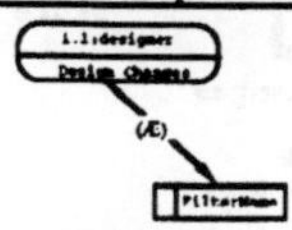	The filter or action is informed of any artefact update (or tool/role) events, rather than enactment events. In the example, FilterName is informed whenever an artefact update event occurs and process stage 1.1 is the current enacted stage (i.e. the stage on which the user is currently working). If the Æ was not present, the filter is only be informed whenever stage 1.1. is enacted or de-enacted.
	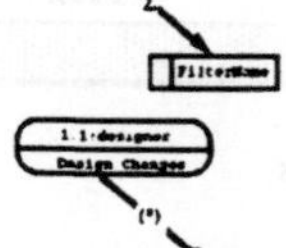	The filter or action is informed when stage 1.1. *or any of its subprocess stages* are enacted or de-enacted. In the example, FilterName is informed whenever an enactment event occurs for stage 1.1 or any of its subprocess stages. If Æ and ∑ are combined, the filter is informed of any artefact update events made when stage 1.1. or any of its subprocess stages are the current enacted stage.
	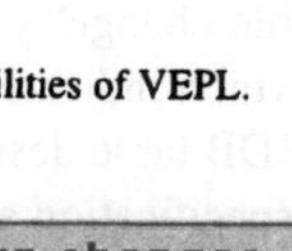	The filter or action is informed of a series of events from the connected process stage (or artefact, tool or role), as opposed to being informed of single events. In the example, several enactment or deenactment events may occur, all of which are sent to FilterName and it will decide on the action it will take based on this sequence of events, not just as each event is received. The * is often combined with ∑ and Æ annotations.

Figure 9. Extended modelling capabilities of VEPL.

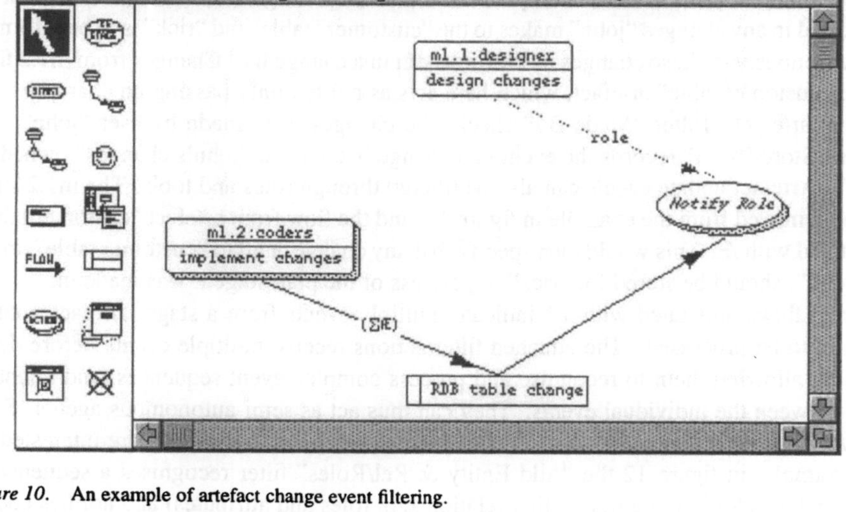

Figure 10. An example of artefact change event filtering.

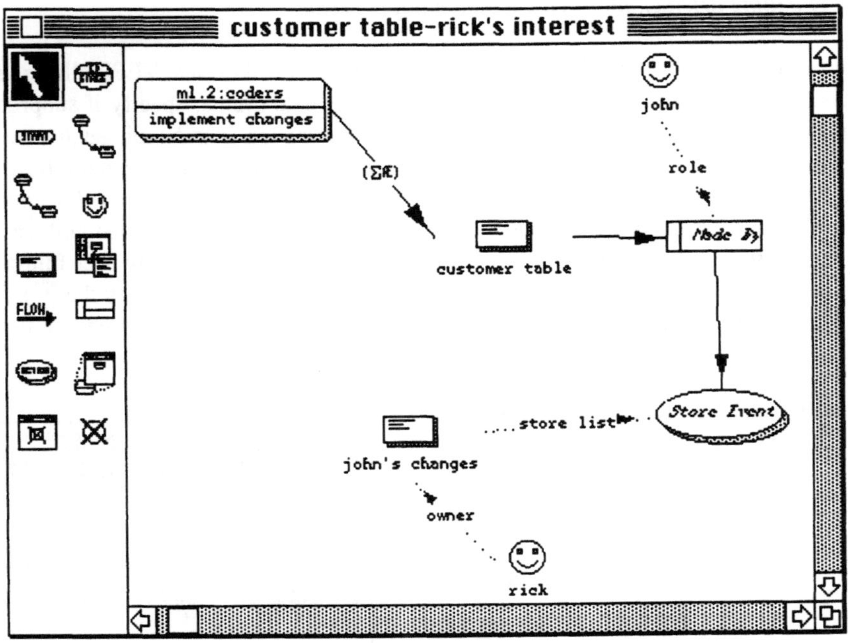

Figure 11. An example of complex filtering and actions.

"designer" is notified of the event. "RDB table change" can be defined in two ways: as a large list of artefact change event patterns which indicate the change is to a table, or by determining if the event originated from the RDB table designer tool.

Roles, tools and artefacts used in a VEPL specification act as filters, if events flow into them from other process model components. For example, figure 11 specifies that "rick" is interested in any changes "john" makes to the "customer" table, and "rick" is to be informed asynchronously of these changes by storing them in a change list. Changes from m1.2 flow to the "customer table" artefact, which here acts as a filter, only passing on change events for this artefact. Filter "Made By" checks the changes were made by user "john", and action "Store Event" records the event in a change list, named "john's changes" owned by "rick". Artefact update events can also be filtered through roles and tools. The m1.2 stage can be removed from the example in figure 11, and the flow from artefact "customer table" annotated with Æ. This would then specify that any change at all to "customer table", made by "john", should be stored for "rick", regardless of the plan stage it was made in.

Event flows annotated with a * indicate multiple events from a stage, artefact, tool or role are to be processed. The attached filter/actions receive multiple events before deactivation, allowing them to recognise and process complex event sequences, and maintain state between the individual events. They can thus act as semi-autonomous agents. Such filter/actions reset their state after receiving the sequence of events they are interested in. For example, in figure 12 the "Add Entity & Rel/Roles" filter recognises a sequence of ER updates (addition of a new entity, relationship, roles and attributes) and notifies coders

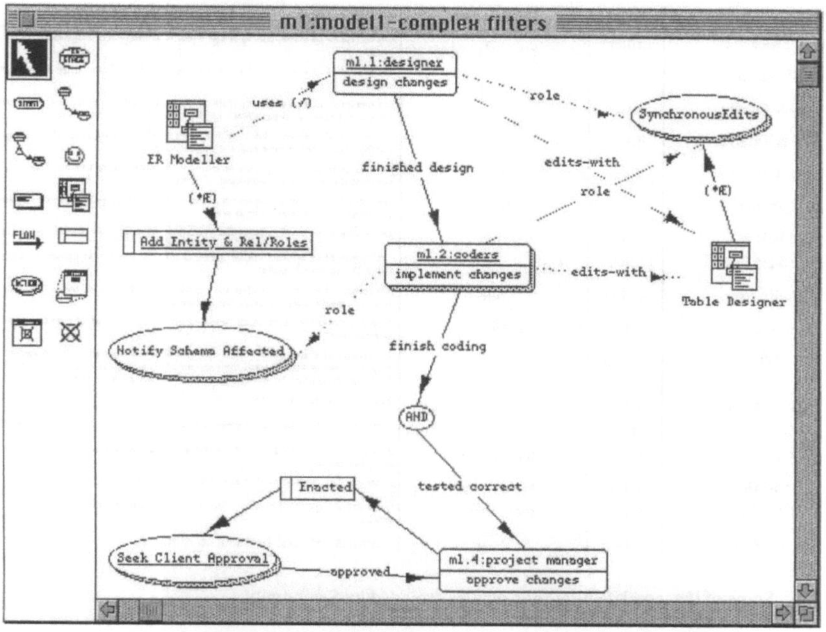

Figure 12. Multiple event processing, stage coordination and external process interfacing.

of the entire table update using a single abstract message, rather than via a long list of messages describing each discrete change made. In this case, the filter accepts ER modeller artefact update events until it recognises an unrelated artefact (e.g., a different table) is being modified. It then passes on a single event to "Notify Schema Affected" and resets its state.

Another example of a complex filter/action is the "SynchronousEdits" action, used in figure 12 to specify how work is to be coordinated between people filling the roles for two concurrently enacted stages. In this case, if the m1.1 and m1.2 stages are concurrently enacted and the roles involved both use the "RDB Table Designer" tool, they must use the synchronous editing mode of that tool to ensure schema changes they are making do not conflict. The "Seek Client Approval" action models a non-computerised process, asking the "project manager" to liaise with the customer and gain approval for the new system changes made.

4.3. Defining filters and actions

Some of the filters and actions shown in the preceeding examples are reusable library templates, some are user-defined by simple specification of patterns to match, some are defined by subprocess models, and others utilise an API interface to the implementation language of Serendipity. Examples of commonly used template filters and actions are shown in figure 13. Users can extend the library by adding their own filters and actions.

Template Filter/Action	Kind	Parameters	Description
Enacted	filter		On receipt of an event, checks if this event is a "start" (i.e. stage enactment) event.
Deenacted	filter		On receipt of an event, checks if this event is a "finish" (i.e. stage deenactment) event.
Made Current	filter		On receipt of an event, checks if this event is a "made_current" (i.e. stage made the current enacted stage) event.
Finished Current	filter		On receipt of an event, checks if this event is a "finished_current" (i.e. stage finished being the current enacted stage) event.
Not Complete	filter	stage	On receipt of an event, passes on event if the given stage has not been completed (i.e. stage's "completed" flag not set).
Complete	filter	stage	On receipt of an event, passes on event if the given stage has been completed (i.e. stage's "completed" flag is set).
Made By	filter	role	On receipt of an event, checks the event was caused by the specified role(s).
Edited By	filter	tool	On receipt of an artefact update event, checks the event was caused by editing using the specified tool(s).
Notify Role	action	role	On receipt of an event, notifies the people filling the role of the connected process stage(s) that the event has occured.
Store Event	action	update list	Stores the event received in an "updates list" artefact (a list of event record descriptions).
Wait	action		On receipt of event, queues the event and does not forward it to connected stages, filters or actions until the user-specified period to wait has completed. Multiple events can be queued using the * annotation on the event flow.
LaunchApp	action	tool	Launches the specified tool.
CreateDoc	action	tool, artefact	Instructs the specified tool to create a new artefact of the specified type & name.
OpenDoc	action	tool, artefact	Launches the specified tool (if not already running) and instructs it to open the specified artefact(s).
SaveDoc	action	tool, artefact	Instructs the specified tool to save the specified artefact(s).
CloseDoc	action	tool, artefact	Instructs the specified tool to close (i.e. stop working with) the specified artefact.
QuitApp	action	tool	Instructs the specified tool to quit.

Figure 13. Some of the template filters and actions provided by Serendipity.

VEPL filter/actions can be defined using VEPL itself by expanding a subprocess, parameterised by start and finish stages, similar to EVPL stage subprocesses. The filter or action icon is then connected to appropriate process model components, and when it receives an event its subprocess is used to determine the response to the event. Figure 14 is a generic filter/action "update data model", which checks if a change to a given "entity" artefact is an entity rename or attribute modification (add, delete, rename or change type), and, if so, stores the artefact change event in an "updates list" specified. The finish stage "update" represents an output from this filter/action, allowing the event to be flowed onto other filters/actions. This filter/action is reused by adding an action icon "update data model" to

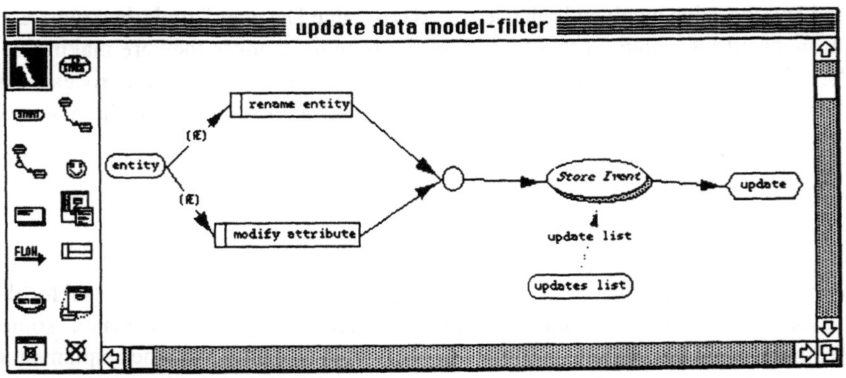

Figure 14. A parameterised filter/action.

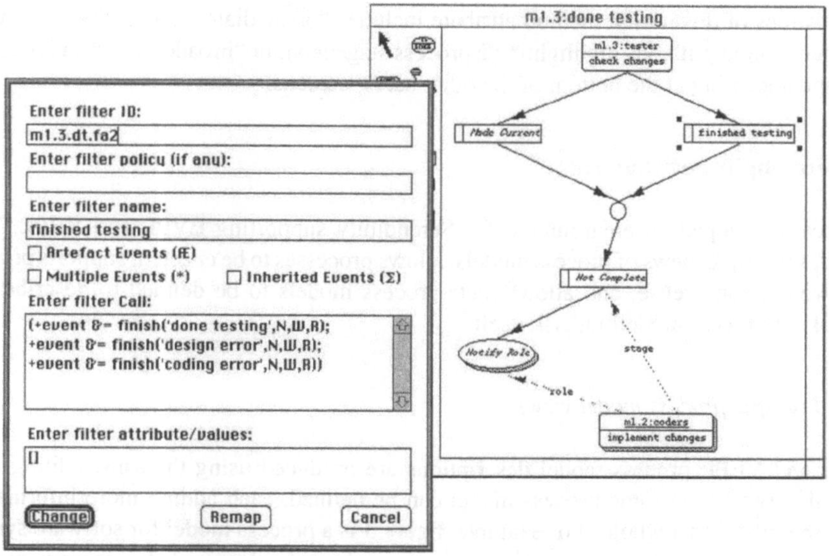

Figure 15. Example of specifying detailed filter and action information.

a process model view. The user then specifies an event flow from an entity artefact into "update data model", and a usage flow into "update data model" from an "updates list" arterfact. The event flow out of "update data model" can optionally be used, as required.

Filters and actions can also be implemented via a Prolog-based API to Serendipity's implementation. Pattern matching on input event representations is used to differentiate the event types. Using the API, filters or actions may call internal functions or access the internal data structures of Serendipity to implement complex filtering operations, or actions.

Figure 15 shows the specification of filter "finished testing". Names and ids are specified in the top three fields. Flags are set to specify the types of event the filter is interested in. The filter call field is either blank (if the filter action is copied from a template or has a subprocess model), or contains Prolog calls. The "&=" operator used in "finished testing" compares an enactment event received by the filter (indicated by "+event") with the given pattern. In this example, the filter responds to finish events in the states 'done testing' 'design error', or 'coding error'. Various other operators and values can be used when specifying the filter/action API call(s), such as the $= operator which compares artefact events to a pattern, the +self value (the object ID of the filter/action receiving the event), and the +from value (the object ID of the model component the event is from). Any parameter inputs to the filter/action specified by usage connections are referred to by +UsageFlowName, where UsageFlowName is specified in a dialog for the usage connection when it is added.

The bottom edit box ("Enter filter attribute/values:") is used to specify state information for a filter/action. This is usually used to constrain or modify filter/action behaviour. For example, for the "Notify Role" action, this has the value "(notify_method=message)", specifying that for this instantiation of the Notify Role template action, coders should be notified by a message (using an e-mail-like messaging system described in Section 5).

Other values of this notify_method attribute include: "open_dialog", to notify users via a dialogue; "highlight", via highlighting a process stage icon; or "broadcast", via a broadcast message appearing at the bottom of the other users' screens.

5. Serendipity environment

We have developed an environment for Serendipity supporting EVPL and VEPL. This provides multiple views of process models, allows processes to be enacted, supports process improvement and reuse, and allows meta-process models to be defined to describe and control the process modelling task itself.

5.1. Multiple process model views

EVPL and VEPL process model descriptions are produced using the same editing tool. Several views of the same process model can be defined, each adding more information about the model as a whole. For example, figure 3 is a process model for software system enhancement, figure 6 describes the artefacts, tools and roles used this process model, and the figures in the previous section associate filters/actions with these stages, artefacts, tools and roles. Textual views of process stages are also available (figure 6). Serendipity keeps all views of a process model consistent under change, or at least, in known state of inconsistency (Grundy et al., 1995a).

5.2. Process enactment

A Serendipity process model can be enacted by users for a particular project at any time. Enacted processes can be modified at any time to extend and improve the process specification, more precisely define the work to be done, or rewrite the history of work done to assist in process improvement and work documentation.

A process model is enacted by selecting the stage to enact and selecting a menu option. The user also specifies a reason for the enactment and, if the stage has subprocess models defined, a start stage to commence in. An enactment event is then sent to the process stage, and start stage (if specified). The start stage immediately sends enactment events to all stages connected to it. As work on a stage is completed or a subprocess completes (a finish event flows into a finish stage), finish events flow into connected stage(s), enacting them. AND stages wait for all stages which flow into them to complete; OR stages enact all stages they flow into upon receipt of a finished event from any input stages. Users can manually complete enacted stages (if no actions prevent this) by supplying a finish state for them.

Figure 16 shows an enacted process model. Enacted stages (i.e., those started and not yet finished) are shaded. Event flow(s) which have activated a stage are also highlighted. The user's current enacted stage (the stage that user is currently working on) is highlighted bold. Users can select any enacted stage, they fill the role of their to be current enacted stage. Users can view a list of their enacted stages, also shown in the figure, and this can function as a "to-do" list for that user.

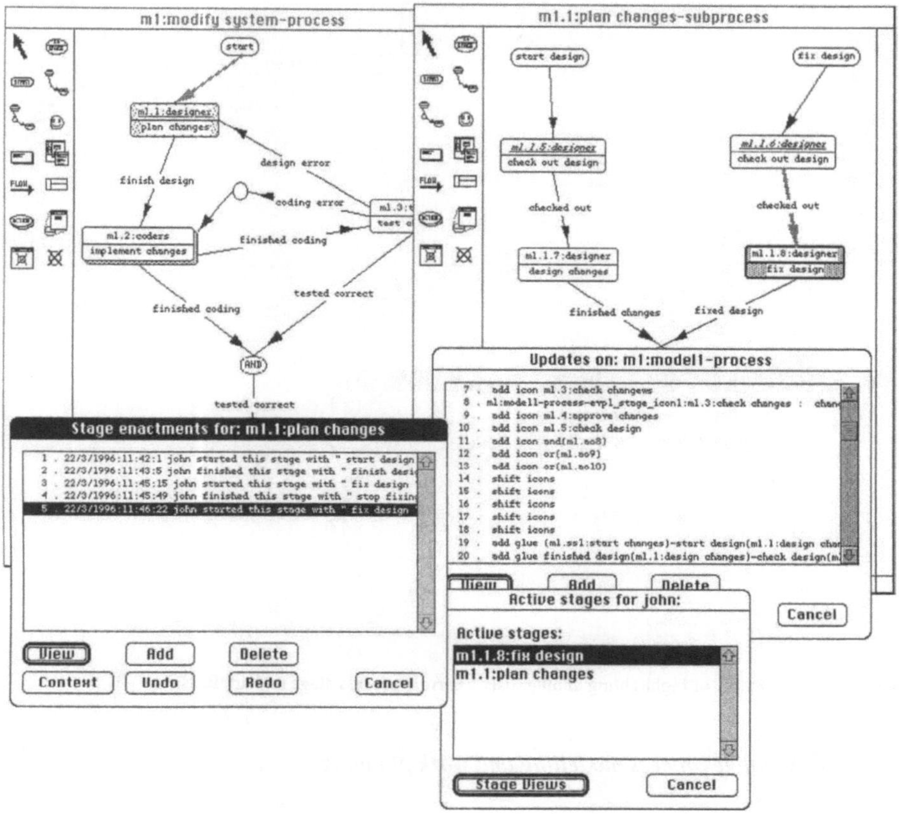

Figure 16. Enactment and modification histories for a stage, and a user's to-do list.

When a stage is enacted, finished, or made/unmade current, it records this event in an enactment history. A reason is also stored specifying when and why the event occurred and which user or stage caused the event. A modification history describing changes to each stage and view is also stored, allowing users to review the evolution of process models. Figure 16 illustrates these two histories for stage m1.1.

"EXCEPTION" actions can be defined for process models. These have no input event flows, but are automatically invoked when a stage, action or filter finishes in a state not handled by the process model i.e., no output event flows match the finishing state name. This usually indicates an error in the process model. "EXCEPTION" actions reduce the complexity of process models by eliminating many error-handling actions and event flows. Multiple "EXCEPTION" actions can be defined for a process model with filters used to determine which one handles a particular exception. If not defined for a model a default handler informs the user of a process model exception (via a change description in a dialog). The user can then handle the exception manually, or extend the model to take the exception into account.

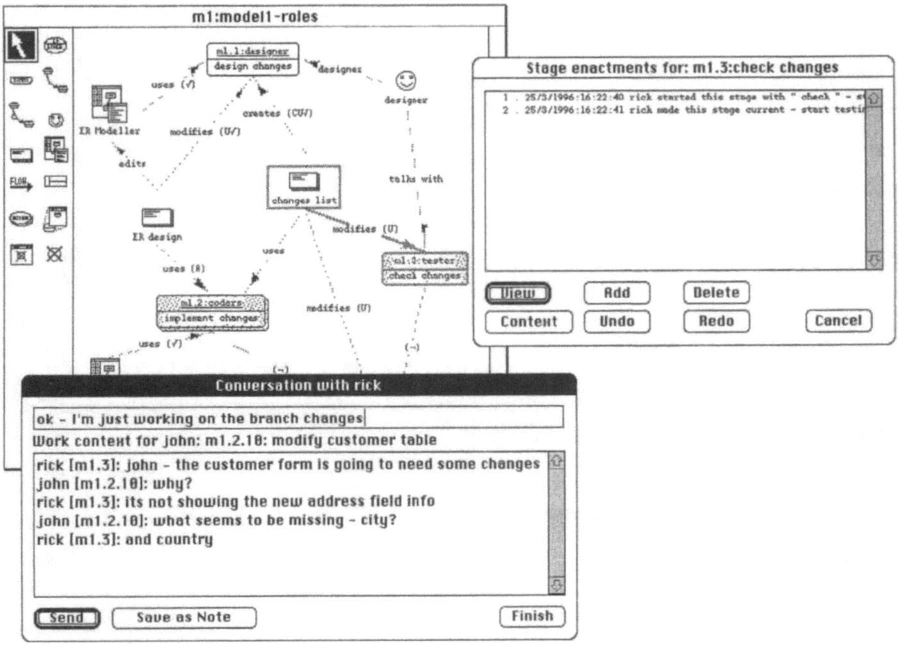

Figure 17. An example of highlighting another user's current process stage and artefacts.

5.3. Collaborative process modelling and work planning

Serendipity process model views may be shared amongst developers, and synchronously, semi-synchronously or asynchronously edited. Serendipity has also been integrated with several small CSCW tools which provide work context-dependent notes, messaging and text chats to facilitate inter-person communication (Grundy et al., 1996b).

To collaborate effectively on a large project, users need higher-level awareness support in addition to the low-level messaging, communication and view editing mechanisms (Grundy et al., 1996b). Serendipity provides information about the work contexts of their collaborators in a high level, synchronous way as shown in figure 17, by highlighting and colouring collaborators enacted stages. Actions can also be specified to highlight artefacts currently being modified by other users, such as the "changes list" in figure 17. Users may have more than one EVPL view open, with enacted and current enacted stages highlighted.

5.4. Process improvement, reuse and meta-process models

Process models often need modification during or after use, as exceptions or unforseen events arise, and may evolve as users become more proficient at describing their work processes, or because details of a process used varies between projects. Serendipity allows process models to be modified at any time: before, during or after use.

To aid in model evolution, Serendipity can provide statistics about the complexity (number of substages, event/usage flows, etc.) of subprocess models, the number of enactments a stage or subprocess model has had, the amount of time each stage has been the current enacted stage, the number of modifications made to models, and the number of work artefact changes made while a stage or its subprocesses were the current enacted stage (Grundy et al., 1996b).

Process stages or filter/actions can be abstracted into process templates. Users can view and edit template models but templates cannot, however, be enacted. Templates are abstract and must first be instantiated (copied) by adding a stage to a process model, specifying it is based on a template.

To guide the process modelling, enactment and improvement process itself, meta-process models can be defined as EVPL and VEPL models. These are also useful for coordinating the modelling activity between multiple collaborators. An added advantage is that changes made to process models can be recorded against the current enacted stage for the meta-process model, allowing collaborators to track reasons for process model changes. Meta-process models also allow users to specify event handling and rules for the EVPL and VEPL notations themselves, by specifying filters and actions on artefact events for process models (i.e., handling events from process model artefacts and tools). Examples of such event handling/rules include specifying ownership and access privileges for process models, controlling when and how process models can be updated, and specifying default exception handling approaches.

6. Using Serendipity with other tools

Serendipity provides more than a process modelling and management environment. Used with other tools, it specifies the *context* of work for multiple collaborators, facilitating activities such as collaborative software development, method engineering and office automation.

A user's current enacted stage defines the work context for that user. Any work artefact changes made using other tools are assumed by Serendipity to be part of the work associated with this stage. Changing the current enacted stage changes the context of work. To illustrate the benefits of this, we briefly describe how Serendipity provides a process modelling tool for an integrated Information Systems Engineering Environment (ISEE) (Grundy et al., 1996c). This environment includes the SPE object-oriented software development environment (Grundy et al., 1995b), an ER modelling tool (MViewsER) (Grundy and Venable, 1995a), a form/report building tool (MViewsDP) (Grundy et al., 1996d), and a NIAM modelling tool (MViewsNIAM) (Venable and Grundy, 1995).

Serendipity does more than just define the work context for a user. As shown in Section 5, stages record their enactment and modification histories. They also record the work artefact updates made while the stage is the current enacted stage, thus documenting the history of work for this work context. Figure 18 shows an example of changes made to an MViewsER view stored for the "m1.2.10: modify customer table" stage for coder "john".

In addition to recording work artefact updates in Serendipity stages, work artefact change descriptions stored by the tools themselves are augmented with information about the work context they were made in. Users of these tools can review work artefact modification

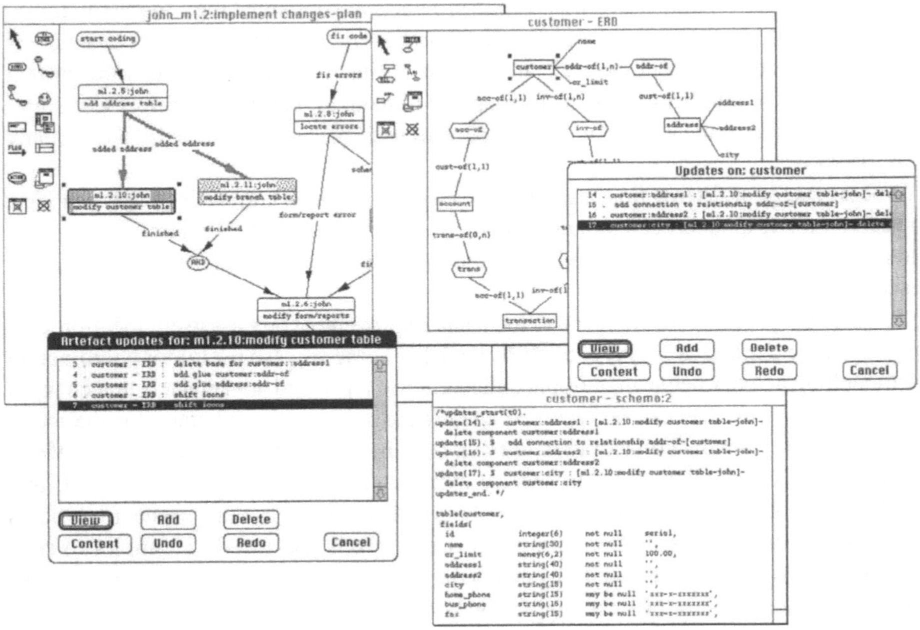

Figure 18. A work artefact change history in Serendipity and work artefact modification history in MViewsER.

histories to determine the changes made, who made them, and when and why they were made (i.e., the work context they were made in). Figure 18 shows an example of a modification history for the "customer" entity/schema from the MViewsER modelling tool. The augmented change descriptions for "customer" are also shown in its schema view, "customer - schema". Note that the changes shown in the "Updates on: customer" dialog and "customer - schema" view are customer artefact updates. Those shown in the "Artefact updates for: m1.2.10: modify customer table" are changes made to the "customer - ERD" view. The user can access the artefact-level updates made in response to these view changes via the "View" button.

While Serendipity highlights process stages, usage connections and artefact and tool icons in use by another developer, as shown in figure 17, it also allows actions to be specified which highlight work artefacts in integrated tool views (Grundy et al., 1996b). Filter/actions defined by a developer can detect work artefact changes made in an integrated tool by someone else, and send messages to the tool of the developer to highlight these icons. Due to this degree of integration of Serendipity and our ISEE, users can also undo or redo work artefact changes within Serendipity or the tools which generated them. Users can also select a work artefact change and request a list of all views of the artefact in the available tool(s). In a tool, users can request a list of views of the appropriate work context of a selected change description.

Serendipity actions can be used to constrain some of the integrated tools' functionality, facilitating "Method Engineering" within our integrated environment. Most software development notations and methodologies are designed to be generic across all problem domains. Research indicates, however, that configuring notations and methodologies to the needs of the particular project results in more appropriate tools and techniques (Harmsen and Brinkkemper, 1995). Computer-aided Method Engineering (CAME) tools support this by allowing system developers to specify which notations (or parts of notations) and methodologies (or parts of methodologies) are to be used for a particular project (Harmsen and Brinkkemper, 1995). We have used Serendipity to support Method Engineering for our ISEE (Grundy et al., 1996c). Serendipity EVPL views specify tool usage and process models ("methodology steps"). VEPL views specify rules and event handling which restrict the use of certain tools and guide or enforce the methodology processes.

Serendipity has also been used to support process modelling for a suite of office automation programs, including Macintosh versions of Microsoft Word^{TM}, Microsoft Excel^{TM}, the GlobalFax^{TM} fax/OCR application, and the Eudora^{TM} e-mail utility.

7. Architecture and implementation

Serendipity and the tools that we have integrated to make up the ISEE described in Section 6 are implemented using the MViews architecture for developing ISDEs. Integration of Serendipty and these tools, although all independently developed, has proved very successful, due to the component and event-based nature of MViews. The integration between Serendipity and the Office Automation tools was more limited, due to the more limited interface provided by these tools.

7.1. MViews

MViews is a framework of object-oriented classes which provides abstractions for implementing ISDEs (Grundy and Hosking, 1993, 1996; Grundy et al., 1996d). New environments are constructed by specialising classes to describe the ISDE repository and view representations. Software system data is described by a graph-based structure, with graph *components* (nodes) specifying e.g., classes, entities, attributes and methods, and *relationships* (edges) linking these components to form the system structure. Multiple views of this repository are defined using the same graph-based structure. These views are rendered and manipulated in concrete textual and graphical forms. External tools built without using MViews can be interfaced to the integrated environment by using "external view" data and event translators. Figure 19 shows an example of the MViewsER Entity-Relationship and relational schema modelling tool developed using MViews (Grundy et al., 1995b). The repository describes entities, relationships and attributes, entity-relationship connections, and schema text. The multiple views provided by MViewsER include graphical ER views and textual schema and documentation views.

MViews uses an event-based software architecture. This supports inter-component consistency management by generating, propagating and responding to *change descriptions* whenever a component is modified. A change description documents the exact change in the state of a component. It is propagated to all relationships the component participates in. Receiving relationships can respond to a change description by applying operations to themselves or other components, forwarding the change description to related components,

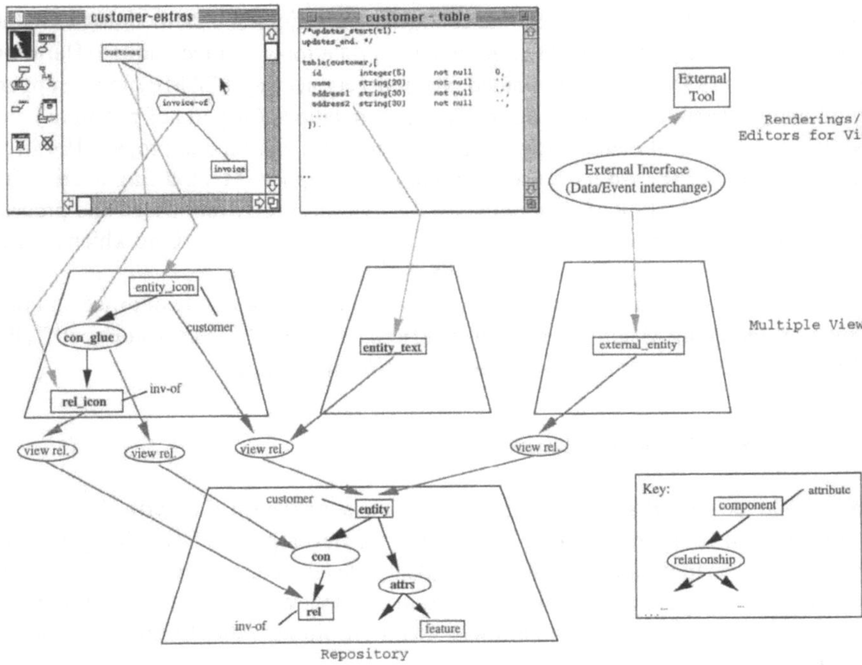

Figure 19. Example of implementing MViewsER using MViews.

or ignoring the change. This technique supports a wide variety of consistency management facilities used by ISDE environments, including multiple view consistency, inter-component constraints, efficient incremental attribute recalculation, undo/redo, and version control and collaborative facilities (Grundy et al., 1996d).

MViews is implemented in Snart, an object-oriented extension to Prolog. Environment implementers specialise Snart classes to define new environment data dictionaries, multiple views, and view renderings and editors (Grundy et al., 1996d). Snart is a persistent language, with repository and view objects dynamically saved and loaded to a persistent object store.

7.2. *Individual modelling tools*

We have built many ISDEs using MViews including SPE (Snart Programming Environment), MViewsER, MViewsNIAM and MViewsDP, the tools integrated into the ISEE. Integration of these tools was achieved using integrated, hierarchical repositories (Grundy and Venable, 1995a, 1995b). This involves either defining integrated repositories which represent, for example, OOA, NIAM and ER data in a common representation, or linking individual repository items with MViews inter-repository relationships, such as between MViewsER schema fields and MViewsDP form components. Inter-repository links keep data in different repositories consistent under change in a similar manner to the view relationships described above.

7.3. *Serendipity environment*

Serendipity was built by specialising MViews classes for representing repository components and relationships, view icons, glue and textual components, and view editors. Constructing EVPL and VEPL diagrams using Serendipity view editors results in construction of repository-level components and relationships describing a process model.

When a process model is used, process model component state variables are modified to indicate receipt of enactment events. Enactment events are simply represented as MViews change descriptions. These are propagated along event connections, with stages interpreting event change descriptions they receive (AND, OR, and start/stop stages interpret these in a special way). Filters and actions interpret change descriptions by comparing the event change descriptions to a filter pattern, running a Prolog predicate which accesses MViews data (i.e., the API interface), or enacting a subprocess model.

Process stage enactment and modification histories are provided by MViews. Artefact modification histories are constructed as stages receive artefact change descriptions from other MViews environments, or from components which interface to applications built without using MViews, and which translate external tool events into MViews change descriptions.

7.4. *Integrating Serendipity with other environments*

Figure 20 shows the way Serendipity interfaces with other MViews ISDEs and external tools. If a change is made to a view-level ISDE tool item (1), this is propagated to a repository ("base view") level item change, and the change description generated by this repository-level item change propagated to the ISDE base view component (2). The base view adds artefact, tool and user information to the change description, and broadcasting of this modified change description is detected by the Serendipity environment base view (3), which then forwards the change description to the user's current enacted stage (4). This, in turn stores the change in its artefact modification history (5) and forwards it to any filters/actions it is linked to by artefact update event connections (6). The process

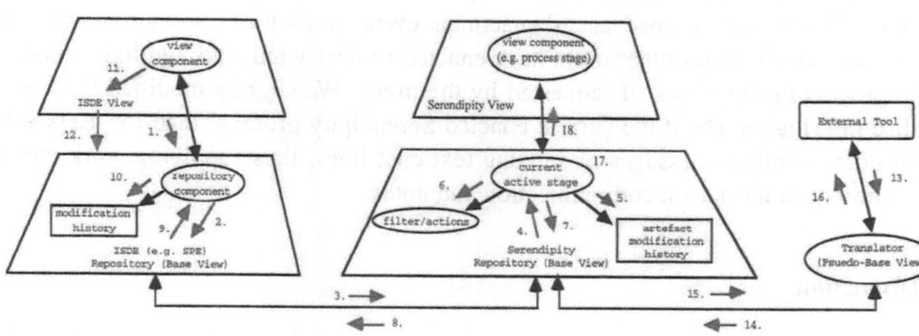

Figure 20. Integrating Serendipity with integrated tool views.

stage attaches "work context" information to the artefact change description (primarily information about the stage itself), sends it back to the Serendipity base view (7), which returns it to the ISDE tool base view (8). The ISDE base view returns the augmented change description to the work artefact which generated it (9), which stores it in its own modification history (10). View updates are sent to Serendipity in the same manner, with the updated view component sending a change description to the view (11), which forwards it to its base view (12), with the base view forwarding the change description to Serendipity and receiving an augmented change description as before. Any change to the enactment status of Serendipity base process model components (17) is detected by their view components (18), which are rerendered to reflect the change.

External tools, such as Microsoft Word™ or Excel™, can also use Serendipity to provide a process modelling and work context environment. Events from such external tools (Apple Events) are sent to a small translator program (13), which forwards them to Serendipity for processing (14). Serendipity events can be sent to the external tools via the translator (15, 16). These might request the external tool be launched, ask it to save, open or close files, or ask it to carry out some other task (such as send a fax or an e-mail message).

Serendipity's collaborative editing facilities are provided by the C-MViews extensions to MViews (Grundy et al., 1995c). Filters and actions specified on process models work as expected if synchronous editing is used as all collaborators share the same process model data. Asynchronous editing implies collaborators have different process model versions which they edit and enact independently, with other collaborators not being aware of these events. Semi-synchronous editing propagates editing, enactment and artefact update events between the environments of collaborators who express interest in these changes for particular process stages, tools, artefacts or roles. C-MViews base views have been extended to incorporate a notion of *monitors* which detect events of interest to collaborating users. Users request, via the C-MViews server, other collaborators' environments to establish monitors on items of interest. Events are then propagated semi-synchronously, via the server, to the interested user's environment for presentation and/or actioning.

The CSCW messaging and note annotation tools provided by Serendipity are separately developed tools which can be used with any MViews environment. The context-sensitive messages store information about the MViews component(s) they are related to, and notes are connected to specified components by relationships. The high-level highlighting of Serendipity process model stages enacted by other users is implemented by having the central C-MViews server broadcast all enactment events to each user's environment. This then records which stages other users have enacted or deenacted, and highlights these on visible process model views (if requested by the user). We slightly modified these tools to utilise information about the current enacted Serendipity process stage for users when storing notes, sending messages or sending text chat lines, thus supplying work context information for inter-person communication and notes.

8. Discussion

Our work has made four main contributions to Process Technology. EVPL provides an expressive, graphical process modelling and work planning language, which can be animated to support both high-level and low-level work context awareness during cooperative work. VEPL is a novel, graphical event handling language that is both accessible for simple use by inexperienced process modellers, and yet expressive enough for experienced process modellers and environment implementers to build sophisticated event-handling systems. The Serendipity environment provides multiple view support for building, enacting, reusing and improving EVPL and VEPL process models, and includes a range of cooperative work capabilities. Our integration of Serendipity with other tools, particularly software development and CSCW tools, provides novel user interface capabilities, with Serendipity allowing EVPL and VEPL to utilise events from these integrated tools in various ways.

8.1. *Experience with Serendipity*

Serendipity has been applied to a range of small to medium process modelling, work planning and coordination, and tool integration problems. The authors have built a number of process models describing: software processes; coordination of collaborative software development; coordination of use of disparate office automation applications; general work processes for academics, students and business people; and for meta-models for method engineering and process improvement. Serendipity has also been used by colleagues and students to describe: software processes; steps in different Information Systems methodologies; and general work processes. The largest process model so far developed has over 200 process stages, artefacts, tools and roles, with over 60 process model and filter/action views. A project is currently in progress to more formally compare Serendipity's languages to other workflow and process modelling languages and to evaluate the Serendipity environment facilities against those of other workflow management systems and process-centred environments.

Serendipity environment performance over the above range of tasks has generally been good, but with some deficiencies, elaborated below. The asynchronous and semi-synchronous CSCW tools and editing capabilities are quite usable in our current implementation, but the synchronous tools have prohibitive performance problems (Grundy et al., 1995c). Processing stages with several filters and actions also cause performance problems, particularly with hierarchical filters ($\sum$ annotation on event flows), which must receive events from all subprocess components. Users can request that filters and actions be actioned semi-synchronously and that the environment process filters for specified stages in idle time. These techniques improve performance with the drawback that enforcement strategies are applied some time after enactments or artefact changes have been made, necessitating occasional roll-back of artefact changes.

Integration of Serendipity with other, independently developed, MViews tools has been successful, producing tightly integrated environments. Serendipity actions can send events to other MViews tools to perform almost any operation, and any tool event can be detected and acted upon by Serendipity filters. This is due to the component and event-based architecture of MViews systems, and the common implementation language and view and repository representation techniques. Integration with heterogeneous tools implemented without MViews has resulted in more limited forms of integration. Serendipity actions can be defined to communicate with these tools but for significant levels of integration, this requires extensive coding and still results in less than complete integration. Serendipity currently runs on the Macintosh only. While we have integrated it in a limited way with a number of commonly available Macintosh applications, it is much more difficult to communicate with applications on other platforms.

8.2. Comparison to other languages and systems

Process modelling languages need to support the description of stages in the process, the data used by process model stages, and the handling of events related to stages, work artefacts and tools. Many process modelling languages use only textual languages to codify tasks and their data. Examples include those of Marvel (Heineman et al., 1992), Adele (Belkhatir et al., 1994), ConversationBuilder (Kaplan et al., 1992a, 1992b), Oz (Ben-Shaul and Kaiser, 1994a) and EPOS (Jaccheri and Conradi, 1993; Conradi et al., 1994). Work artefacts are usually coded as types, for example in EPOS these are "dataentity" specialisations and in Adele they are aggregate type relations. Steps in a process model are usually encoded as rules, for example in EPOS as task types and Adele as event triggers. While these languages allow users to precisely specify process model data and activities, they are difficult for many users to understand and modify (Baldi et al., 1994; Bogia and Kaplan, 1995; Swenson et al., 1994), and cannot readily be utilised for high-level work context awareness, such as via animation. Our EVPL and VEPL process modelling languages have the same expressive power as these textual languages, but use a primarily graphical approach to visualising process model activities and data (with detailed process stage, artefact, tool and role characteristics specified in forms and textual views). EVPL diagrams provide significantly clearer representations of process model structure than their textual counterparts, as process modellers can see multiple levels of abstraction using EVPL diagrams. VEPL provides a graphical approach to expressing event handling, allowing process modellers to better visualise event flow and handling than a comparative textual encoding.

A variety of graphical process modelling languages have been developed for use by process-centred environments and workflow management systems. Examples include E^3 p-draw's E^3 PML (Baldi et al., 1994), SPADE's SLANG (Bandinelli et al., 1993, 1994, 1996), ProcessWEAVER's transition nets (Fernström, 1993), Action Workflow's loops (Medina-Mora et al., 1992), Regatta and TeamFLOW's Visual Planning Language (VPL) (Swenson, 1993), and wOrlds's obligation nets (Bogia and Kaplan, 1995). Graphical modelling capabilities of some of these systems, such as VPL, SLANG and wOrlds, is limited to specification of the stages in a process and the "enactment flow" between these stages. For example, SLANG encodes activity structure as a form of petri-net, but textually encodes activity artefacts and tools as process types and transitions as guards and actions; VPL represents process stages graphically with interconnecting flows, but lacks representations of stage data and enactment rules; and wOrlds represents aspects of social structure graphically, including obligations between "process stages", but provides only a simplistic graphical representation of process data and obligation rules. Our work, and work with E^3 PML, ProcessWEAVER and Action Workflow, has found graphical modelling of process stage data (artefacts, tools and roles), and the visualisation data shared by multiple process stages, to be very important when using complex process models. E^3 PML uses an object-oriented process modelling notation, with a variety of representations (stages, artefacts, tools), and interconnections (task decomposition, inheritance, control flow). E^3 PML represents task decomposition on the same diagram, whereas EVPL utilises multiple overlapping and hierarchical views, making decomposition easier to manage and resulting in clearer process models. E^3 PML also lacks the range of usage connections and annotations of EVPL, making it less expressive. ProcessWEAVER (Fernström, 1993) uses a transition net (a form of petri-net) to model cooperative procedures via tasks and subtasks, with token flow driving enacted process advancement. The transition nets do not explicitly represent artefacts, tools and roles, but nodes can be defined which correspond to their manipulation. ProcessWEAVER nets thus do not make clear the roles involved in stages of a process, nor the artefacts and tools, nor do they indicate usage of this work context information, as in EVPL. Action Workflow (Medina-Mora et al., 1992) uses a document flow model to represent work processes, which has been found to be deficient for many process modelling situations (Swenson et al., 1994). EVPL uses a more flexible and expressive state-transition model, as do VPL, ProcessWEAVER and SPADE, with EVPL processes driven by the propagation of enactment events. This notion of explicit enactment events in EVPL, rather than state transition by token propagation in SPADE and ProcessWEAVER or the use of speech acts in VPL, is important, as it is utilised by VEPL to handle enactment, artefact update and tool events in a homogeneous fashion.

The event handling of most graphical process modelling languages is codified graphically for "enactment" events (state transitions), but textually for stage guards and actions, and for coding interaction with people or other tools. Most workflow management systems, such as Action Workflow, Regatta and TeamFLOW, provide a limited range of "interesting events", supporting interaction with other tools and some forms of notification and work coordination based on event occurrence. These are usually codified using a form-based approach where modellers specify simple actions to carry out based on a range of possible events. VEPL provides a more powerful, clearer and extensible event-handling language than this approach. VEPL filter/actions can handle more complex events, provide a wider range of actions, be extended by building hierarchical models (or using an API), and have graphical representations which more clearly show event flow in a process model. Most process modelling languages use textual rules with guard predicates and executable actions to specify how events are handled. ProcessWEAVER provides a textual co-shell language which allows users to specify actions for process model nodes when fired by input tokens. SLANG uses textual specifications of guards and actions for nodes in a state transition network. Marvel and Oz use guarded rules specified over data types. Adele provides an activity manager, which uses a textual language to specify database-related event handling for process models. All these approaches use textual languages, and hence complex rules suffer from the same problems as when presenting other process model structures textually. The relationships between different stages, tools and artefacts are expressed in a linear, textual fashion, unlike graphical VEPL event handling models where event flows and process model artefacts are more clearly represented in graph form. This is a particular problem for rules which span multiple process stages, as understanding rule behaviour requires several textual data and task specifications to be viewed. VEPL permits specification of both simple and very complex event handling graphically, resulting in high-level visualisations of event handling behaviour. For example, the specification of a simple work coordination mechanism where a user is notified whenever another user modifies a particular work artefact, shown in figure 11, results in an intuitive event handling model which is relatively easy for users to understand and modify. By allowing previously defined filter/action templates to be composed using the same event propagation notation, VEPL supports reuse of a wide range of event handling behaviour. In contrast, approaches which use textual codification to handle these kinds of events, produce event handlers that are less clear and generally less reusable than those of VEPL. Visual dataflow-based languages, such as Fabrik (Ingalls et al., 1988) and Prograph (Cox et al., 1989), provide graphical dataflow models which are similar in nature to VEPL, but use dataflow rather than event-flow, which is less appropriate in a process modelling domain. Some visual languages, such as ViTABaL (Grundy and Hosking, 1995), utilise an event-driven model but lack the equivalent of Serendipity's filters, actions, and interest specification capabilities. Because of their general-purpose nature, these visual programming languages lack specific process modelling capabilities, and thus cannot express and represent process model event-handling and work coordination tasks as effectively as EVPL and VEPL.

Most process modelling languages are supported by their own PCE or Workflow Management System (WFMS). For example E^3 PML is supported by the E^3 p-draw tool (Jaccheri and Gai, 1992), SLANG is supported by the SPADE PCE (Bandinelli et al., 1993, 1994, 1996), VPL by the Regatta (Swenson et al., 1994) and TeamFLOW (TeamWARE, 1996) WFMSs, and obligation nets by the wOrlds CSCW environment (Bogia and Kaplan, 1995). These environments typically allow process models to be enacted and advanced as users complete work on tasks and subtasks. Environments utilising textual languages are limited in how they can inform users of the state of enacted process models. Most avoid using the textual codification of the process model to visualise their state. In contrast, EVPL process models are animated to give users feedback on process model enactment and other users' work. Action Workflow, wOrlds, Regatta and TeamFlow also use highlighting of enacted process models. However in Serendipity, we have gone further than these systems, and highlight stages as they are enacted/deenacted, highlight stages of cooperating users, and highlight artefacts, tools and roles currently being used by an enacted process stage (Grundy et al., 1996b).

Most recent Computer Supported Cooperative Work (CSCW) research has focused on low-level interaction mechanisms, such as synchronous and asynchronous editing. Examples include most Groupware systems (Ellis et al., 1991), GroupKit (Roseman and Greenberg, 1996), Mjølner (Magnusson et al., 1993), CMViews (Grundy et al., 1995c), and Rendezvous (Hill et al., 1994). These systems lack information about the "work context" changes have been carried out in, whereas Serendipity makes this information readily accessible to users. There is some work on providing higher-level process modelling and coordination facilities, such as workflow configuration (Medina-Mora et al., 1992), obligations (Kaplan et al., 1992b; Bogia and Kaplan, 1995), and shared workspace awareness (Roseman and Greenberg, 1996), but these systems are generally separate from the work artefacts or editing tools, and are not used to provide work context information. In Serendipity the text chats and note annotation CSCW tools incorporate information from the enacted process models, as do other tools integrated with Serendipity. This provides users with more context awareness capabilities than are supported by other CSCW and process modelling systems.

Oz (Ben-Shaul and Kaiser, 1994a) enables the definition of high-level work coordination capabilities, such as summits and treaties, utilising extensions to the Marvel rule-based model (Heineman et al., 1992), and supports cooperative transactions for itself and integrated tools (such as ProcessWEAVER) via an external concurrency control architecture (Heineman and Kaiser, 1995). TeamFLOW, wOrlds and ConversationBuilder use obligations between process stages to enable enactment events to be propagated between process stages from different diagrams. ProcessWEAVER (Fernström, 1993) uses transition networks to model synchronisation of concurrent activities by cooperative agents, but like wOrlds these deal mainly with "enactment"-style events. VEPL supports the definition of complex work coordination by using EVPL constructs in its event-handling models. We have used VEPL to build a variety of low-level concurrency control mechanisms, typically supported by groupware tools. We have also used it to build higher-level work coordination models similar to those used by ProcessWEAVER and wOrlds (Grundy et al., 1996b). Serendipity uses both a shared server to broadcast events and share data, and a repository storing local process model and work artefacts. This allows it to support Oz-style summits, with process model enactment events generated by multiple users broadcast via the shared server. VEPL models can handle and constrain data sharing, in a similar manner to the "diplomats" utilised by Oz treaties, and can also support concurrent transactions by having actions store sequences of artefact changes and undo them if multiple transaction conflicts arise.

There have been several attempts in integrating CSCW, ISDEs and workflow/PCEs, including ConversationBuilder (Kaplan et al., 1992b), MultiviewMerlin (Marlin et al., 1993), wOrlds (Bogia and Kaplan, 1995), SPADE/ImagineDesk (Di Nitto and Fuggetta, 1995; Bandinelli et al., 1996), ProcessWEAVER and Oz (Heineman and Kaiser, 1995), and various groupware tools and Oz (Ben-Shaul et al., 1994b; Ben-Shaul and Kaiser, 1996). Many of these, such as MultiviewMerlin, SPADE/ImagineDesk and wOrlds, have produced environments with a high degree of integration. However none have produced environments which capture detailed information about enacted process models and present this "work context" information using the integrated tools themselves, nor do they allow information from the integrated tools to be utilised in the process modelling environment exactly as though the tools were part of that environment. In contrast, in our integrated environment the boundaries of the two systems, from a user's perspective, have disappeared, despite there being no modifications to the underlying architectures of either system. The augmentation of SPE and CSCW tool data with Serendipity process model information, and the highlighting of in-use SPE artefacts by Serendipity, allows users to remain aware of when and why other users have made artefact changes or caused tool events to occur. Similarly, SPE artefacts (classes, diagrams etc.) and tools (OOA/D editor, text editor, debugger etc.) can be represented in Serendipity as EVPL artefacts and tools, and VEPL filter/actions can use these artefacts and tools as both producers of events and as event filters. Serendipity process stages can also store work histories using the event descriptions generated by the integrated tools. This provides closer user interface integration between process modelling tool, CSCW tools and work tools than do these other PCE, CSCW environment and work tool integration efforts.

Our integration of Serendipity, MViews CSCW tools, and an ISEE has been very successful, due to the event-driven and component-based natures of these environments and their common implementation platform. We have achieved a much lesser degree of integration of Serendipity with third-party tools, because of limitations in the interfaces provided to the tool data and control functions. Recent work on integrating such heterogeneous software development, CSCW and process modelling environments has focused on data aspects (e.g. federated approaches to tool integration (Bounab and Godart, 1995)), control integration via enveloping (e.g. between Oz and third-party tools (Valetto and Kaiser, 1995)), and process integration (e.g. coordinating tool usage via PCEs (Marlin et al., 1993; Di Nitto and Fuggetta, 1995; Bandinelli et al., 1996)). Serendipity provides, via MViews and VEPL, an event-driven interface which supports control and process integration with disparate tools. At present, Serendipity utilises a repository which stores process model data in the form of Prolog terms. While Serendipity actions can be defined to facilitate import/export of this data with heterogeneous tools, this does involve more work than utilised in federated approaches to tool integration via distributed data management (Bounab and Godart, 1995; Valetto and Kaiser, 1995).

8.3. *Current and future work with Serendipity*

The experience of the authors and other users of Serendipity to date has been generally positive, but a number of desirable improvements to the languages and environment have been identified. The EVPL and VEPL languages have proved concise and yet expressive for a large range of process modelling and work coordination/event-handling problems. There is a need, however, to support better artefact structuring and repository querying facilities for users. To this end we are currently developing a third visual notation for Serendipity, the Visual Query Language (VQL). This will allow users to specify the structure of artefacts using an MViews-style component and relationship model. Users will also be able to define complex, graphical queries over MViews (and other tool) environment repositories. Triggering of filters and actions will then be possible over the query result, which will be incrementally updated as data is modified.

We have been working on filters and actions which allow work contexts to be determined dynamically, in addition to being specified using artefact, role and tool representations. This will allow Serendipity to determine changes of context (i.e., change of current enacted process stage) automatically, rather than requiring users to do this manually, as at present. Actions will utilise the history of work context information for stages to determine such context changes. Visualisation techniques allowing users to view summaries of inter-related changes made in different contexts are also being developed. These facilities will, we believe, assist Serendipity in better supporting more informal aspects of work, which is currently poorly supported by existing workflow and PCEs (Kaplan et al., 1996).

We are currently porting Serendipity to Java and are utilising a Web-based interface to the modelling language views (Grundy et al., 1997). This will improve the portability of our environment. More significantly, this will enable us to provide better integration with heterogeneous tools, or at least utilise the now commonly-understood "plug-in" and "helper application" model of loose integration employed by browser-style applications. Another advantage will be the ability to reuse other people's CSCW and work tools, in the form of Java applets, plug-ins or helper applications. We have found that systems built on event-based architectures, such as those of MViews, are much more amenable to tool integration, and plan to use Serendipity's event-handling language in conjunction with such systems to better facilitate their integration with Serendipity.

9. Summary

Our work with Serendipity makes a number of new contributions to research into process technology. EVPL, while having some similarities to other graphical process modelling languages, utilises a versatile enactment event-based execution model. EVPL is used for generic process modelling, meta-process modelling and detailed work planning, and supports process stage work context modelling via representations of artefacts, tools, role communication and various kinds of inter-component usage relationships. VEPL is a novel event-handling notation which utilises EVPL model components, as well as introducing filters and actions, to handle enactment, artefact update, tool-induced and role-induced events. EVPL and VEPL together allow process modellers to specify arbitrary event-handling for process models, which includes specifying a large variety of work coordination strategies, automatic process model rule application, and propagation and storage of events between stages, artefacts, tools and roles.

Serendipity provides a support environment for EVPL and VEPL which includes multiple views of process models and event handling specifications. Enactment of process models utilises highlighting of enacted stages, event flows and artefact, tool and role representations, helping multiple users to keep aware of others' work contexts. Collaborative editing of process models is supported, along with CSCW tools for annotating, messaging and talking about models. Integration of Serendipity and MViews-based tools for performing work results in environments with highly integrated user interfaces. Descriptions of changes made in other tools are annotated with work context (i.e., process stage) information, process stages store lists of changes made while they are enacted (forming histories of work), and icons in both Serendipity and the integrated tools are highlighted to facilitate group awareness. Such tight user interface integration is not supported by most other PCE/workflow and work tool integration efforts. Serendipity utilises a component, event-based architecture. A hybrid technique is used to store process model information in local repositories for speed of access, and a central repository used to enable sharing of versions and broadcasting of enactment, artefact and tool events.

Serendipity is being extended to incorporate a flexible Visual Query Language, which will double as a tool repository query and data visualisation language, and a more flexible artefact data specification for use by VEPL filters and actions. We are currently porting MViews to Java and will also port Serendipity and the MViews suite of software development tools, to improve their accessibility, performance, and ability to integrate existing tools. Filters and actions are currently interpreted, but will be compiled to Java to improve their performance.

Acknowledgments

The authors gratefully acknowledge the many helpful comments of the anonymous reviewers on earlier drafts of this paper.

References

Baldi, M., Gai, S., Jaccheri, M.L., and Lago, P. 1994. Object oriented software process design in E^3. In *Software Process Modelling & Technology*, A. Finkelstein and J. Kramer, and B. Nuseibeh (Eds.), Research Studies Press.

Bandinelli, S., Fuggetta, A., and Ghezzi, C. 1993. Software process model evolution in the SPADE environment. *IEEE Transactions on Software Engineering*, 19(12):1128–1144.

Bandinelli, S., Fuggetta, A., Ghezzi, C., and Lavazza, L. 1994. SPADE: An environment for software process analysis, design and enactment. In *Software Process Modelling & Technology*, A. Finkelstein, J. Kramer, and B. Nuseibeh (Eds.), Research Studies Press.

Bandinelli, S., Di Nitto, E., and Fuggetta, A. 1996. Supporting cooperation in the SPADE-1 environment. *IEEE Transactions on Software Engineering*, 22(12).

Barghouti, N.S. 1992. Supporting cooperation in the marvel process-centred SDE. *Proceedings of the 1992 ACM Symposium on Software Development Environments*, ACM Press, pp. 21–31.

Belkhatir, N., Estublier, J., and Melo, W.L. 1994. The adele/tempo experience. In *Software Process Modelling & Technology*, A. Finkelstein, J. Kramer, and B. Nuseibeh (Eds.), Research Studies Press.

Ben-Shaul, I.Z. and Kaiser, G.E. 1994. A paradigm for decentralized process modeling and its realization in the Oz environment. *Sixteenth International Conference on Software Engineering*, IEEE CS Press, pp. 179–188.

Ben-Shaul, I.Z., Heineman, G.T., Popovich, S.S., Skopp, P.D., Tong, A.Z., and Valetto, G. 1994. Integrating groupware and process technologies in the Oz environment. *9th International Software Process Workshop: The Role of Humans in the Process*, C. Ghezzi (Ed.), Airlie, VA: IEEE CS Press, pp. 114–116.

Ben-Shaul, I.Z. and Kaiser, G.E. 1996. Integrating groupware activities into workflow management systems. *7th Israeli Conference on Computer Based Systems and Software Engineering*, Tel Aviv, Israel, pp. 140–149.

Bogia, D.P. and Kaplan, S.M. 1995. Flexibility and control for dynamic workflows in the wOrlds environment. *Proceedings of the Conference on Organisational Computing Systems*, Milpitas, CA: ACM Press.

Bounab, M. and Godart, C. 1995. A federated approach to tool integration. *Proceedings of CAiSE'95*, LNCS 932, Finland: Springer-Verlag, pp. 269–282.

Conradi, R., Hagaseth, M., Larsen, J., Nguyen, M.N., Munch, B.P., Westby, P.H., Zhu, W., and Liu, C. 1994. EPOS: Object oriented cooperative process modeling. In *Software Process Modelling & Technology*, A. Finkelstein, J. Kramer, and B. Nuseibeh (Eds.), Research Studies Press.

Cox, P.T., Giles, F.R., and Pietrzykowski, T. 1989. Prograph: A step towards liberating programming from textual conditioning. *Proceedings of the 1989 IEEE Workshop on Visual Languages*, IEEE Computer Society Press, pp. 150–156.

Di Nitto, E. and Fuggetta, A. 1995. Integrating process technology and CSCW. *Proceedings of IV European Workshop on Software Process Technology*, LNCS, Leiden, The Nederlands: Springer-Verlag.

Ellis, C.A., Gibbs, S.J., and Rein, G.L. 1991. Groupware: Some issues and experiences. *Communications of the ACM*, 34(1):38–58.

Fernström, C. 1993. ProcessWEAVER: Adding process support to UNIX. *2nd International Conference on the Software Process: Continuous Software Process Improvement*, Berlin, Germany: IEEE CS Press, pp. 12–26.

Grundy, J.C. and Hosking, J.G. 1993. A framework for building visual programming environments. *Proceedings of the 1993 IEEE Symposium on Visual Languages*, IEEE Computer Society Press, pp. 220–224.

Grundy, J.C. and Hosking, J.G. 1995. ViTABaL: A visual language supporting design by tool abstraction. *Proceedings of the 1995 IEEE Symposium on Visual Languages*, Darmsdart, Germany: IEEE CS Press, pp. 53–60.

Grundy, J.C. and Venable, J.R. 1995a. Providing integrated support for multiple development notations. *Proceedings of CAiSE'95*, LNCS 932, Finland: Springer-Verlag, pp. 255–268.

Grundy, J.C. and Venable, J.R. 1995b. Developing CASE tools that support integrated design notations. *Proceedings of the 6th European Workshop on Next Generation of CASE Tools*, Finland, pp. 109–116.

Grundy, J.C., Mugridge, W.B., Hosking, J.G., and Apperley, M.D. 1995a. Coordinating, capturing and presenting work contexts in CSCW systems. *Proceedings of OZCHI'95*, Wollongong, Australia, pp. 146–151.

Grundy, J.C., Hosking, J.G., Fenwick, S., and Mugridge, W.B. 1995b. Connecting the pieces. In *Visual Object-Oriented Programming*, M. Burnett, A. Goldberg, and T. Lewis (Eds.), Manning/Prentice-Hall.

Grundy, J.C., Mugridge, W.B., Hosking, J.G., and Amor, R. 1995c. Support for collaborative. Integrated software development. *Proceeding of the 7th Conference on Software Engineering Environments*, IEEE CS Press, pp. 84–94.

Grundy, J.C. and Hosking, J.G. 1996. Constructing integrated software development environments with MViews. *International Journal of Applied Software Technology*, 2(3-4):133–160.

Grundy, J.C., Hosking, J.G., and Mugridge, W.B. 1996a. Towards a unified event-based software architecture. In *Joint Proceedings of the SIGSOFT'96 Workshops*, L. Vidal, A. Finkelstein, G. Spanoudakis, and A.L. Wolf (Eds.), ACM Press, pp. 121–125.

Grundy, J.C., Hosking, J.G., and Mugridge, W.B. 1996b. Low-level and high-level CSCW in the Serendipity process modelling environment. *Proceedings of OZCHI'96*, Hamilton, New Zealand: IEEE CS Press.

Grundy, J.C., Venable, J.R., Hosking, J.G., and Mugridge, W.B. 1996c. Coordinating collaborative work in an integrated information systems engineering environment. *Proceedings of the 7th Workshop on the Next Generation of CASE Tools*, Crete.

Grundy, J.C., Hosking, J.G., and Mugridge, W.B. 1996d. Supporting flexible consistency management via discrete change description propagation. *Software—Practice & Experience*, 26(9):1053–1083.

Grundy, J.C., Mugridge, W.B., and Hosking, J.G. 1996e. A Java-based toolkit for the construction of multi-view editing systems. *Proceedings of the Second Component Users Conference*, Munich.

Grundy, J.C., Hosking, J.G., and Mugridge, W.B. 1997. A visual, Java-based componentware environment for constructing multi-view editing systems. *Proceedings of 2nd Component Users Conference*, Munich.

Harmsen, F. and Brinkkemper, S. 1995. Design and implementation of a method base management system for a situational CASE environment. *Proceedings of the 2nd Asia-Pacific Software Engineering Conference*, Brisbane: IEEE CS Press, pp. 430–438.

Heineman, G.T., Kaiser, G.E., Barghouti, N.S., and Ben-Shaul, I.Z. 1992. Rule chaining in Marvel: Dynamic binding of parameters, *IEEE Expert*, 7(6):26–32.

Heineman, G.T. and Kaiser, G.E. 1995. An architecture for integrating concurrency control into environment frameworks. *Proceedings of the 17th International Conference on Software Engineering*, Seattle, Washington: IEEE CS Press, pp. 305–313.

Hill, R.D., Brinck, T., Rohall, S.L., Patterson, J.F., and Wilner, W. 1994. The rendezvous architecture and language for constructing multi-user applications. *ACM Transactions on Computer-Human Interaction*, 1(2):85–125.

Ingalls, D., Wallace, S., Chow, Y.Y., Ludolph, F., and Doyle, K. 1988. Fabrik: A visual programming environment. *Proceedings of OOPSLA'88*, ACM Press, pp. 176–189.

Jaccheri, M.L. and Gai, S. 1992. Initial requirements for E^3: An environment for experimenting and evolving software processes. *Proceedings of EWSPT'92*, Trondheim, Norway, pp. 99–102.

Jaccheri, M.L. and Conradi, R. 1993. Techniques for process model evolution in EPOS, *IEEE Transactions on Software Engineering: Special Issue on Software Process Evolution*, 19(12):1145–1156.

Kaplan, S.M., Tolone, W.J., Carroll, A.M., Bogia, D.P., and Bignoli, C. 1992a. Supporting collaborative software development with conversationbuilder. *Proceedings of the 1992 ACM Symposium on Software Development Environments*, ACM Press, pp. 11–20.

Kaplan, S.M., Tolone, W.J., Bogia, D.P., and Bignoli, C. 1992b. Flexible, active support for collaborative work with conversationbuilder. *ACM Conference on Computer-Supported Cooperative Work*, ACM Press, pp. 378–385.

Kaplan, S.M., Fitzpatrick, G., Mansfield, T., and Tolone, W.J. 1996. Shooting into orbit. *Proceedings of Oz-CSCW'96*, DSTC Technical Workshop Series, University of Queensland, Brisbane, Australia, pp. 38–48.

Kellner, M.I., Feiler, P.H., Finkelstein, A., Katayama, T., Osterweil, L.J., Penedo, M.H., and Rombach, H.D. 1990. Software process modelling example problem. *Proceedings of the 6th International Software Process Workshop*, Hokkaido, Japan: IEEE CS Press.

Krishnamurthy, B. and Hill, M. 1994. CSCW'94 workshop to explore relationships between research in computer supported cooperative work & software process. *Proceedings of CSCW'94*, ACM Press, pp. 34–35.

Magnusson, B., Asklund, U., and Minör, S. 1993. Fine-grained revision control for collaborative software development. *Proceedings of the 1993 ACM SIGSOFT Conference on Foundations of Software Engineering*, Los Angeles CA, pp. 7–10.

Marlin, C., Peuschel, B., McCarthy, M., and Harvey, J. 1993. MultiView-Merlin: An experiment in tool integration. *Proceedings of the 6th Conference on Software Engineering Environments*, IEEE CS Press.

Medina-Mora, R., Winograd, T., Flores, R., and Flores, F. 1992. The Action Workflow approach to workflow management technology. *Proceedings of CSCW'92*, ACM Press, pp. 281–288.

Roseman, M. and Greenberg, S. 1996. Building real time groupware with groupkit, a groupware toolkit. *ACM Transactions on Computer-Human Interaction*, 3(1):1–37.

Swenson, K.D. 1993. A visual language to describe collaborative work. *Proceedings of the 1993 IEEE Symposium on Visual Languages*, IEEE CS Press, pp. 298–303.

Swenson, K.D., Maxwell, R.J., Matsumoto, T., Saghari, B., and Irwin, K. 1994. A business process environment supporting collaborative planning. *Journal of Collaborative Computing*, 1(1).

TeamWARE, Inc. 1996. *TeamWARE Flow*, http://www.teamware.us.com/products/flow/.

Tolone, W.J., Kaplan, S.M., and Fitzpatrick, G. 1995. Specifying dynamic support for collaborative work within wOrlds. *Proceedings of the 1995 ACM Conference on Organizational Computing Systems*, Milpitas, CA, pp. 55–65.

Valetto, G. and Kaiser, G.E. 1995. Enveloping sophisticated tools into computer-aided software engineering environments. *IEEE Seventh International Workshop on Computer-Aided Software Engineering*, pp. 40–48.

Venable, J.R. and Grundy, J.C. 1995. Integrating and supporting entity relationship and object role models. *Proceedings of the 14th Object-Oriented and Entity Relationship Modelling Conference*, LNCS 1021, Gold Coast, Australia: Springer-Verlag.

Automated Software Engineering 5, 61–96 (1998)

APEL: A Graphical Yet Executable Formalism for Process Modeling

S. DAMI dami@imag.fr
J. ESTUBLIER jacky@imag.fr
M. AMIOUR amiour@imag.fr
LSR, Actimart Bat 8, av. de Vignate, 38610 Gieres, France

Abstract. Software process improvement requires high level formalisms for describing project-specific, organizational and quality aspects. These formalisms must be convenient not only for capture but also for execution purposes. In order to fulfill these requirements and to build a software process environment capable of supporting engineering tasks we have designed a new graphical, but still enactable, formalism called APEL (for Abstract Process Engine Language).

APEL is very ambitious in the sense that it aims to cover a wide spectrum of needs and approaches expressed not only in the software engineering field but also in many others such as real-time systems, object-oriented methodologies, tool integration, CSCW, workflow and information systems. It is then not surprising to see that many concepts and techniques used in APEL are borrowed from these connected fields. A major outcome of the work presented here was to integrate a broad range of concepts and paradigms in a single and coherent framework, but on the basis of a minimal set of primitive concepts which make it very easy to extend. Globally, the aspects which received most attention are: openness, reuse, scalability, human orientation and cooperative work.

1. Introduction

It is increasingly accepted that software quality, cost and productivity are significantly affected by the software production process. The survival of any company lies in its ability to produce goods of high quality at a low price, i.e., to control and improve its processes. Consequently, it has become a top-level priority to promote and support process improvement. And this is the goal of the PERFECT[1] project. One way to address process improvement is to rely upon a high-level executable formalism which supports project performance, measurement and organisation.

Our research revealed that current process support tools and formalisms, including off-the-shelf tools such as Process Weaver (Fernstrom, 1993) and Adele (Estublier and Casallas, 1994; Estublier 1996; Belkhatir et al., 1993) used within the PERFECT project, did not provide a wide enough range of functionality required for large development projects. Our initial attempt to widen the scope of these tools by integrating them, showed that combining their two respective low-level process engines produced yet another low-level process engine, with even more ambiguously related concepts.

We regard high-level tools as those which use concepts that can be mapped directly onto concepts in the users' problem domain, while low-level tools make use of concepts and mechanisms which users use to build the higher-level concepts that they need. Thus, we regard systems that deploy petri nets, rules or triggers to describe process activities, coordination or collaboration as low-level process engines. Experience has shown us that such systems require extensive programming effort and expertise.

Our goal, then, was to define a high level process formalism, i.e., a formalism intuitive enough for non-experts to describe the problem to solve, but precise enough to provide real execution support. However, the risk of defining a high level formalism is to oversimplify, and to end up with a system which is not capable of providing efficient, flexible and comfortable support. Since our goal was to develop an industrial tool (not just a prototype), APEL had to meet the following requirements:

- it must be intuitive and usable by non-experts (high-level),
- it must cover a wide range of process modelling domains (not only software),
- it must be executable and capable of supporting fine grain engineering tasks,
- it must be platform independent.

In order to meet these requirements, we have designed a high-level, but still executable, graphical formalism, called APEL (for Abstract Process Engine Language). APEL is intended to cover a wide range of modeling domains and to be platform-independent. What is meant by "platform-independent" is that it can be (at least partially) implemented using different existing process engines. As a matter of fact APEL is now fully supported by Adele for the configuration management domain, partially supported by Process Weaver for the Workflow domain; it may be, in the future, partially supported by Esterel for the manufacturing domain, and by Catia for the virtual product management domain. We hope that implementations for other domains, will be undertaken in the future.

A prototype support environment on top of Adele has been implemented in 1996; a complete and pre-industrial support environment is currently under-way. Figure 1 shows its overall architecture and how the APEL graphical editor fits in.

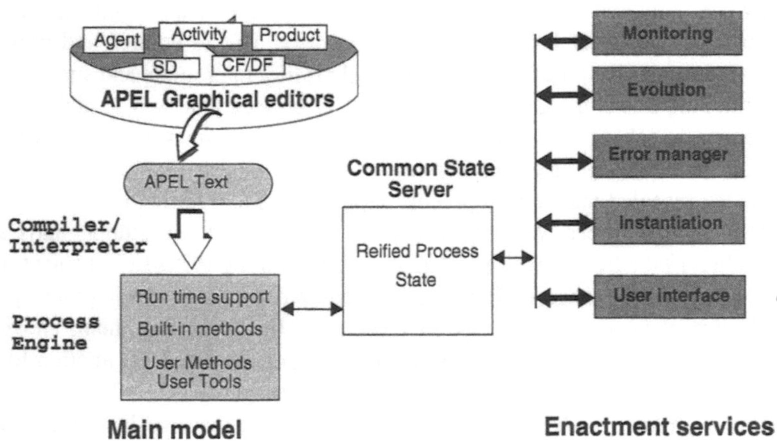

Figure 1. Architecture of the APEL environment.

APEL has two representations: (1) a graphical one, intended for end-users and high-level process description, and (2) a textual one for tools and advanced users. The architecture consists of:

- a graphical editor for capturing and modeling processes,
- a translator from the graphical into the textual representation,
- a compiler from the textual to the executable formalism. Currently, we use the Adele executable formalism, but other process support platforms may be used,
- a common state server that maintains the current state of the running process,
- a set of enactment services ranging from the user interface manager to the monitoring service.

This paper aims at presenting the APEL graphical formalism. Issues about the overall architecture will be addressed in a forthcoming paper.

The paper is structured as follows. Section 2 presents the basic concepts of the APEL formalism. Section 3 deals deeper into its more advanced features and presents how some concepts are integrated, in particular from a reuse perspective. In Section 4, the current approaches used in the process modeling field and other connected fields are discussed together with the approach used for APEL. Finally, in Section 5 we conclude the paper by discussing the current state of the work, some open issues as well as future works.

2. APEL basic concepts

The description of the APEL main concepts is depicted in figure 2.

The top part of figure 2 shows the basic concepts used for APEL, namely *activity*, *product*, and *agent*. These concepts may have *states*, contrary to those of *measure* and *version* which

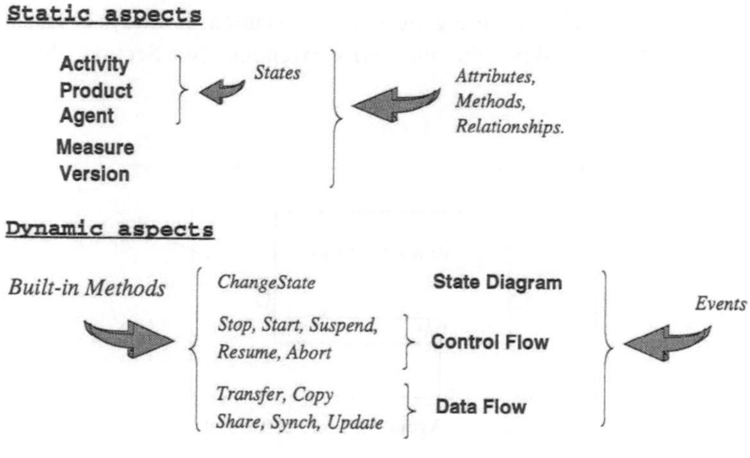

Figure 2. Basic concepts.

are also defined but with which no states are associated. A state can only be changed upon **event** occurrences. The notions of state and event are central to the dynamic aspects of the meta-model including control flow, data flow and the state diagram (SD).

It is worth underlining that our formalism borrows a number of concepts and mechanisms from other fields (real time systems (Harel, 1988a; Selic et al., 1994), Object-oriented methodologies (Rumbaugh et al., 1991; Booch, 1994; Jacobson, 1992; Coad and Yourdon, 1991a, 1991b), tools integration (Purtilo, 1994; Reiss, 1990; Frankel, 1990), CSCW (Malone et al., 1992; Roseman and Greenberg, 1992; Conklin and Begeman, 1988; Paoli and Tisato, 1994; Grudin, 1994; Reinhard et al., 1994), workflow (Georgakopoulos et al., 1995; Gruhn, 1994; Mohan et al., 1995)) and seeks to address a broad-ranging set of domains, integrating notably the workflow dimension. The aspects which have received top-level attention are: openness, human orientation, reuse, scalability and cooperative work.

2.1. *Static aspects*

Static aspects encompass all of the entities involved in an APEL process definition. They are mainly activities, products, and agents as presented in figure 3. In addition to these three basic types of objects, APEL allows one to define measure and version types (presented in Section 3.7).

An activity is an atomic or composite operation, or a step in a process that contributes to the realization of an objective (Promoter Group, 1996). Activities play a central role, because it is both the APEL concept for process structuring, and the cornerstone on which reuse is achieved. A process model is composed of a set of activities which can be recursively decomposed, following different levels of abstraction starting from the generic to the most specific one. The concept of activity is akin to the programming languages concept of module. An activity is defined as the union of an interface and a body. The interface defines the visible part of an activity including inputs/outputs, involved agent/role, conditions that must be satisfied before, after or during the activity execution. It always has a single entry point, a normal exit point and possibly three error exit points (see Section 2.5).

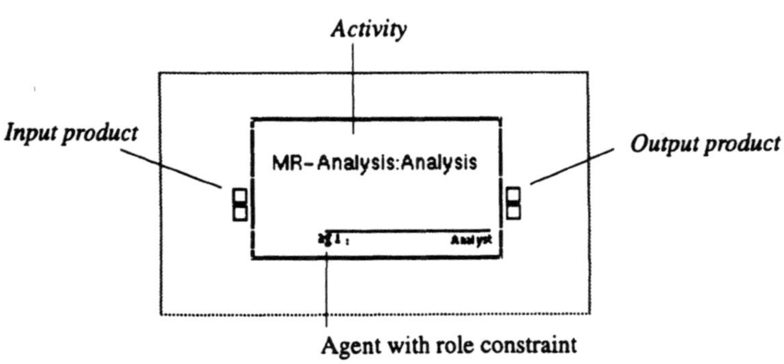

Figure 3. Activity, product and agent concepts[2].

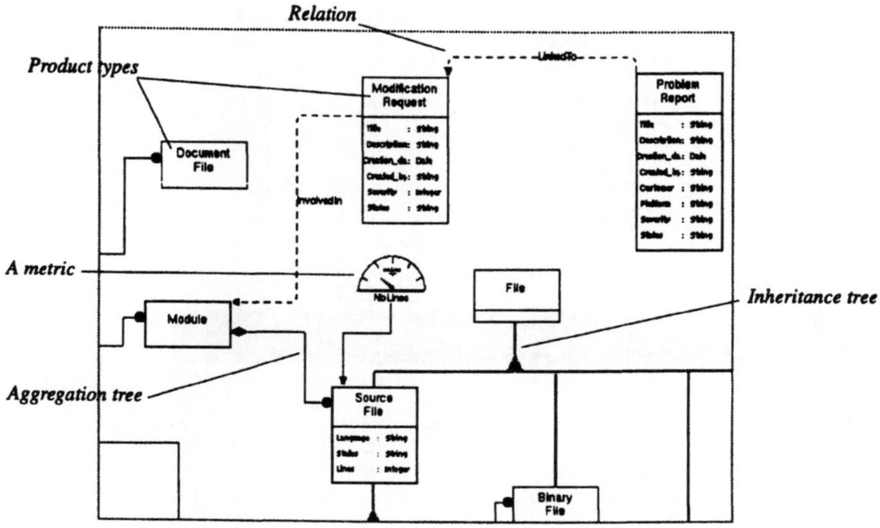

Figure 4. A partial view of a product types diagram.

An activity is represented by a box; by convention, time flows from left to right; thus the left side of the box corresponds to the beginning of the activity, and the right side to its end.

In APEL, any entity (be it a product, a role, an activity or a measure) defined in a process model is typed, and therefore type definitions are supported in the different graphical views (see figure 4). As in conventional object-oriented languages, types are described through attributes and methods, and can be organized into inheritance or aggregation trees. The graphical notation used for types definition is very similar to OMT (Rumbaugh et al., 1991) notation.

2.2. *Dynamic aspects*

In APEL, the dynamic aspects of the process are described through control flow, data flow and the state diagram.

The **control flow** (CF) is the collection of activities and "rules" that allow their execution and ordering in time. It describes aspects dealing with the sequencing of operations and interaction between concurrent activities. Those aspects are concerned with time and changes, in contrast with the static definition of activities.

The **data flow** (DF) shows how products circulate between activities that consume, transform or produce them. Thus, a data flow connects the output of an activity to the input of another activity. It represents an intermediate data state within a process. Normally, the state is not changed by the data flow.

State diagrams (SD) are defined for product, agent and activity types. For a given type, the SD describes (1) how an entity of that type evolves over time (i.e., which states it goes through), and (2) the events and conditions that cause the states changes.

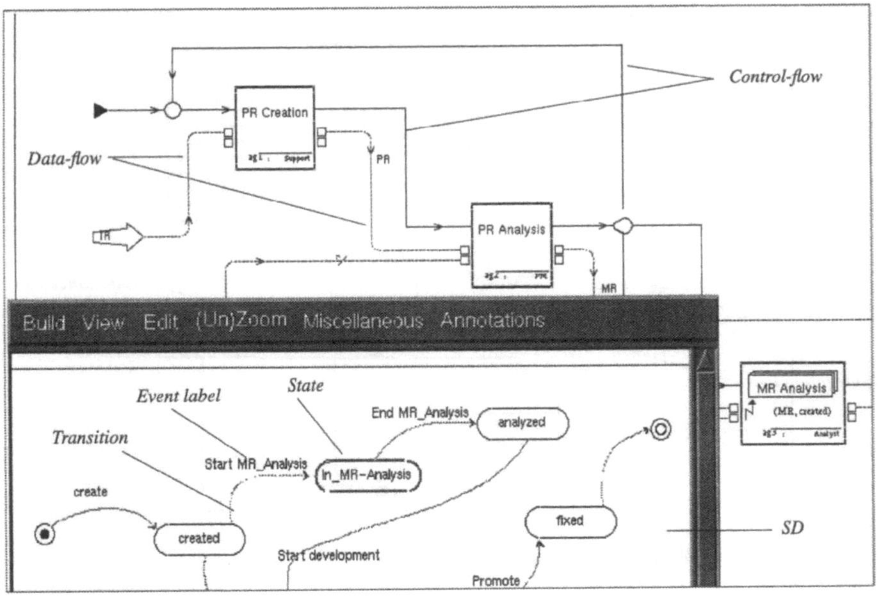

Figure 5. Control flow, data flow and state diagram.

A state change is called a *transition*, and is produced when an event is captured. Thus, an SD may be viewed as a finite state machine with an initial state, a set of final states, a set of possible intermediate states, and a set of possible transitions between states.

Figure 5 is a screen dump showing (1) in the lower left corner, a state transition diagram (see Section 2.4) defined for the data type called MR (for Modification Request). The background is a partial view of the control flow (plain arrows) and data flow (dashed arrows) diagrams (see Sections 2.5 and 2.6).

Note finally that products along with their SDs allow product-centered modeling of the process, while activities with control and data flow, allow activity-centered modeling.

2.3. Events

The dynamic aspects are based on the concept of event and event capture. In APEL events are *generated automatically* each time:

- a *method* is called.
  These events are noted «*m(producer [, activity])*» where *m* is the method, *producer* the object on which the method is executed, and *activity* the activity in which the method is called[3]. If *activity* is not specified, the current activity is retained.
- the state of an entity is changed.
  These events are noted «*(producer, new-state [, activity])*» where *producer* is the object which state has changed, *new-state* the new state of the object, and *activity* (optional) the activity in which the object state has changed.

- a *temporal* (clock) signal is produced.
  These events are noted «*(producer, event-id, 'TMP')*» where *producer* is an entity, *event-id* an identifier and *TMP* a string constant.

All event instances are generated in a systematic way and broadcast. Those which are actually handled must be recognized (we say **captured**) by potential consumers; otherwise they are ignored. An event capture is a way of expressing, at model level, the relationship between the producer of an event, and the consumer. An event capture, or simply a **capture**, is a pair (*event-definition; logical-expression*). An event instance matching the *event-definition* is said to be captured by entity *X* (known as the *consumer* by default) if, when the event occurs, the logical expression is true.

Events are captured by the entities (activity, product, or agent) that need to react to that event. The different dynamic models (SD, CF and DF) only express how the entity reacts when the event is captured:

- In an **SD**, the entity is an instance of any type, and the reaction is to change the state of that entity (by calling the built-in method[4] *ChangeState*).
- In a **CF**, the entity is an activity instance, and the reaction is to change the state of that activity (by calling the built-in methods *ChangeState*, *Start*, *Stop*, *Abort*, *Resume* or *Suspend*).
- In a **DF**, the entity is a product instance, and the reaction is to move data between activities (by calling the built-in methods *Transfer/Copy/Synchronize*, *etc.*).

Consequently, capturing events is the cornerstone of the whole APEL system, with respect to the dynamic behavior and the enaction. This very simple approach is reflected in the textual language which is simply a list of rules of the form:

```
Event-capture => List of methods;
```

All the dynamic aspects are controlled by very few concepts: captured events producing the execution of only 11 built-in methods.

2.4. *State diagrams*

State transition diagrams (SDs) express, for each type, the legal states for the instances of that type, the valid sequence of states (transitions), and the event captures which trigger the transitions from a state to another (see example of figure 6).

To represent SDs we have adopted a notation inspired by Harel's statecharts (Harel, 1987, 1988b; Harel and Gery, 1996), with some adaptations to fit software configuration needs. Statecharts constitute a simple and expressive way to represent, in a modular fashion, the behavior of complex systems. They attempt to overcome the drawbacks of conventional finite state automata, (flat, not structured, complex). Statecharts have been adopted in different domains in computer science, for example, real time systems (Statemate (Harel, 1988a)), and object-oriented design and analysis methodologies (OMT (Rumbaugh et al., 1991), OOD (Booch, 1994)).

In the software process field, SDs have been used to represent the life cycle of objects. For example, in software configuration management, SD is the unique technique used

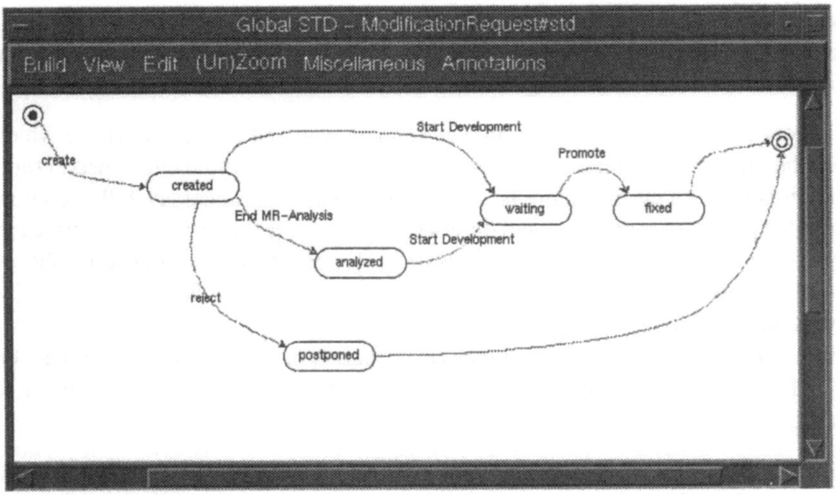

Figure 6. Example of product SD.

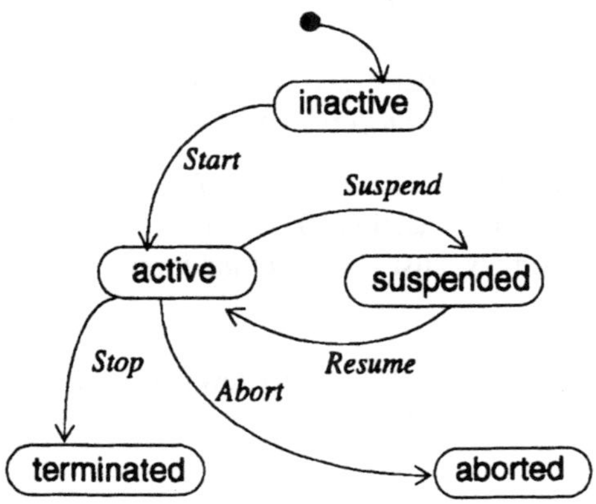

Figure 7. The default SD for activity type A.

to support processes. For that reason, perhaps, SD is representative of the object-centered approaches for software support. In fact, SDs have been used more for modeling procedures and modules' behavior than object life cycles.

Activities have a predefined SD with the states *inactive*, *active*, *suspended*, *terminated* and *aborted* (figure 7). The semantics of *inactive* state is that the activity is waiting (ready) to be launched. The *active* state means that the activity is being executed, while *suspend* means that it has been temporarily interrupted. The *terminated* state means that the activity has been terminated successfully. This state represents the normal exit point of the activity. The semantics of the *aborted* state is that the activity was forced to terminate before completion. This may be the case when an anomaly is detected during the execution of the activity. This state represents an error exit point of the activity.

These states can be reached only by executing the APEL built-in methods *ChangeState*, *Start*, *Stop*, *Suspend*, *Resume* and *Abort*. Conversely, when changing state, an activity *a1*

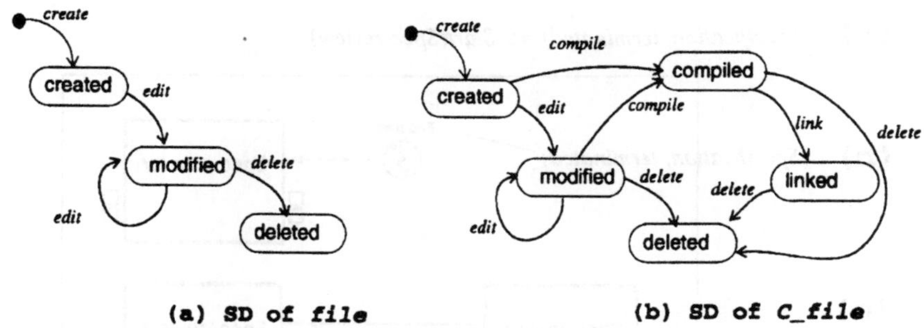

Figure 8. Inheritance of SDs.

of type *A* generates the events (*a1:A*, *active*), (*a1:A*, *terminated*) or (*a1:A*, *aborted*). This seems to be sufficient for defining any control flow.

Calling the method *Start(ReviewSpec)* for the activity *ReviewSpec* of type *Review*, generates the event *Start(ReviewSpec:Review)* which is captured by the SD of *ReviewSpec*, and moves it from state *inactive* to *active*. This state change, in turn, generates the event (*ReviewSpec:Review*, *active*). In the figure above, the simplified event notation *Start* on the arrow is a label used for capturing purpose. The full event specification (i.e., *Start(?:A)*) can be obtained by a simple mouse click.

The SD of a type is inherited by all its subtypes[5]. A subtype can only extend the inherited SD by adding new states and/or new transitions. To illustrate this, we consider the SD of a product type *file* (figure 8(a)) with three states *created*, *modified* and *deleted*, with the transitions shown in the figure, and we define a new product type *C_file* (that represents modules written in C) as a subtype of *file*. In the SD of *C_file* (figure 8(b)) the same states and transitions of *file* are present. Two additional states *compiled* and *linked* are defined with five additional transitions.

2.5. *Control flow*

The control flow diagram describes how activities are to be launched and synchronized during the process execution.

In APEL, the control flow corresponds to the definition and control of the activities' state transitions. Only one of the six built-in methods (*ChangeState*, *Start*, *Suspend*, *Resume*, *Stop*, *Abort*) can change the state of an activity; thus a CF can always be translated into rules that call one of these six methods.

Each one of these six methods can be executed as a consequence of the occurrence of an event. APEL provides a graphical notation for each one of these pairs (event, control flow method).

Figure 9(a) shows how a rule may be represented using the textual language of APEL. The rule means that when the event (*Specification*, *terminated*) occurs (upon the termination of the *Specification* activity) the *Start* method is applied on *Spec-review* to automatically start the activity.

In figure 9(b) the consumer of the event, here the *Spec-review* activity, is represented graphically. In figure 9(c) both the consumer and the producer are represented graphically.

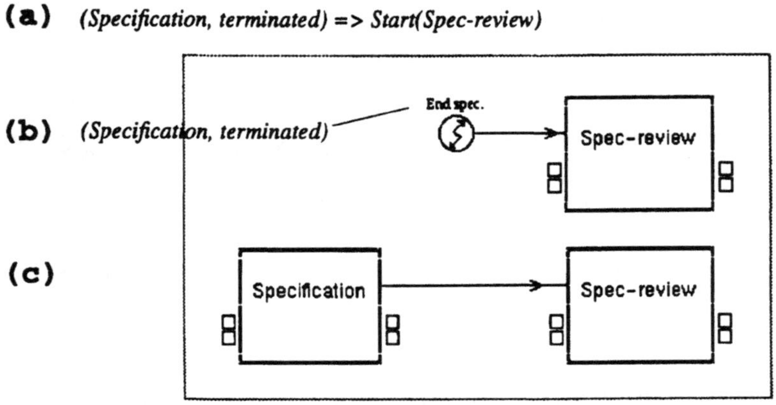

Figure 9. Textual and graphical representations—events and states.

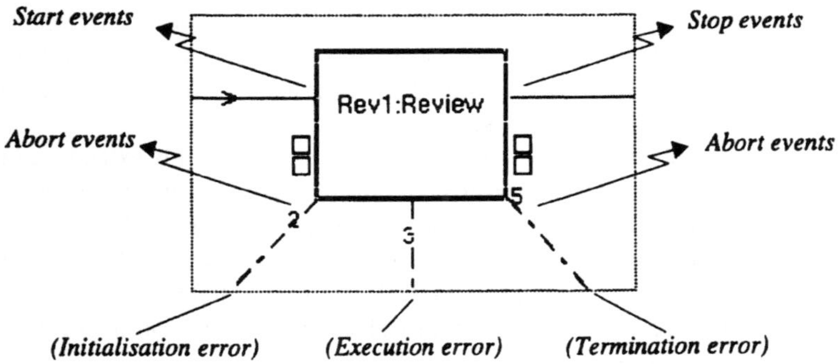

Figure 10. Control flow events.

The two activities are connected by a CF arrow (figure 9(c)) expressing a sequential ordering of the two.

A *Start* event means the start method must be applied to the activity upon capture of that event. *Abort* events may arise for three reasons: (1) when some initial conditions are not satisfied when the activity starts, (2) when some conditions are not satisfied at activity termination, (3) or when an abort is forced during the activity execution (due to external reasons, e.g., a deadline is reached, etc.). Figure 10 shows how *Start*, *Stop* and *Abort* events are represented in a CF.

2.6. *Data flow*

Data flow deals with the data circulation/exchange that may occur between different activities of the process. This covers input/output exchange between activities at their start/end as well as during their execution. Since each activity has its own desktop[6] (one of the functions of a desktop is to maintain and isolate the activity products (see Section 3.1), a transfer may be done toward an *inactive* (or *suspended*) activity (not being executed), in which case, the transferred data is placed on the desktop of the destination activity. The activity will become active if and only if all its input data are received on the desktop and the control flow signal is passed.

We distinguish between **Synchronous** and **Asynchronous** data flow.

In a *synchronous* DF the data is transferred to the destination activity when the source activity terminates. Figure 11 below shows a typical sequential DF. Figure 11(a) shows the formal representation. The *Transfer* built-in method means that the data called *doc* in the *Specification* activity must be moved into *Spec-review* activity under the name *spec*. Figure 11(b) shows the graphical representation of a synchronous DF. The dashed arrow linking the left and right sides of activities means that the data transfer is to be performed at the end of the first one and before the start of the second.

In an *asynchronous* DF the data is transferred between concurrent activities. Two activities can exchange information with no need for them to be sequentialized. In particular, this kind of DF is required for "daemon" activities. Such a data flow is shown (figure 12) by a dashed arrow between two activities, starting from the top or bottom side of the activity box. In the figure, it is specified that the data item *d1* is moved from *A1* to *A2* while *A1* and *A2* are being executed. The transfer occurs when *d1* reaches the state *s* (noted *d1/s*) in activity *A1* (upon the capture of the event (*d1*, *s*, *A1*)). This can be represented formally

(a) *(Specification, terminated) => Transfer(doc, Specification, Spec-review, spec)*

(b)

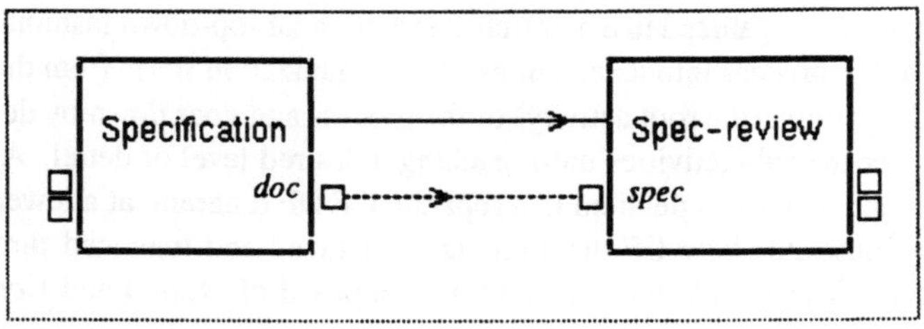

Figure 11. Synchronous data flow representation.

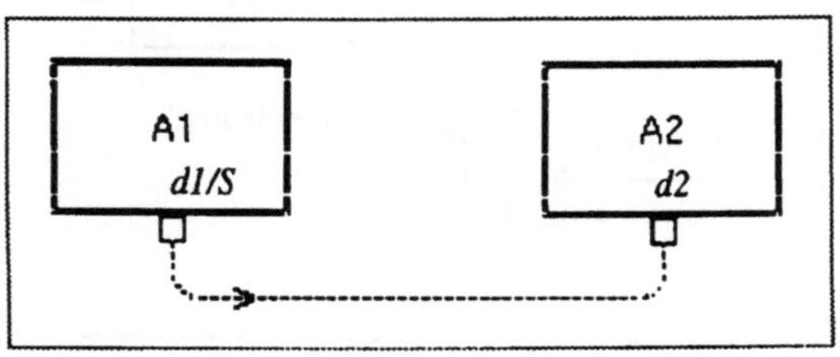

Figure 12. Asynchronous data flow.

by the expression:

```
(d1, s, A1) => Transfer(d1, A1, A2, d2).
```

2.7. *CF/DF connectors*

A connector links different control flows and/or different data flows. The usual control flow connectors are the *And* and *Or* connectors. The *And* connector expresses that the control leaves the connector as soon as all incoming CF signals arrive, while for the *Or* connector, the control leaves as soon as only one of the incoming CF arrives.

CF and DF connectors are special kinds of activities with specific behaviors. Being predefined, and for modeling convenience, these activity types have specific graphical representations that allow type declaration to be dispensed with.

Generally, any useful operator, working on the control or data flow can be coded as specialized activities. This is interesting for applying operators to lists, arrays, sets, records and so on. Consequently, APEL does not provide these operators built-in, but rather, a library of operators compatible with the underlying platform capabilities.

In the example of figure 13, activity *l* is a list operator which delivers as output the different components of the input list *L*. Each element is processed by activity *A*. As soon as *A* terminates, the next item is provided in input and *A* is restarted.

2.8. *Activity decomposition*

Following the decomposition of activities, an APEL process model can be viewed as a stack of CF/DF diagrams organized in a hierarchic and modular top-down fashion that shows the breakdown of the process into its activities. This organization starts from the most general or abstract description (the root activity) of the process and goes down by defining detailed descriptions for the sub-activities until reaching a desired level of detail. At each level of abstraction, each activity is detailed in a separate CF/DF diagram, at a lower level.

Figure 14 illustrates how CF diagrams are organized and managed through activities decomposition. The *Development* activity is composed of *Design* and *Coding* activities.

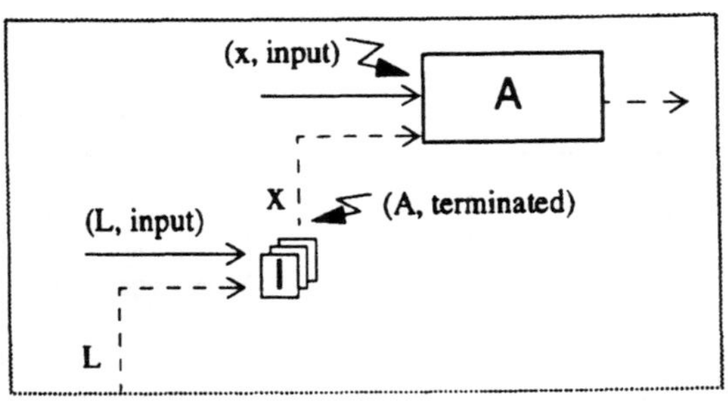

Figure 13. Implementing a list operator.

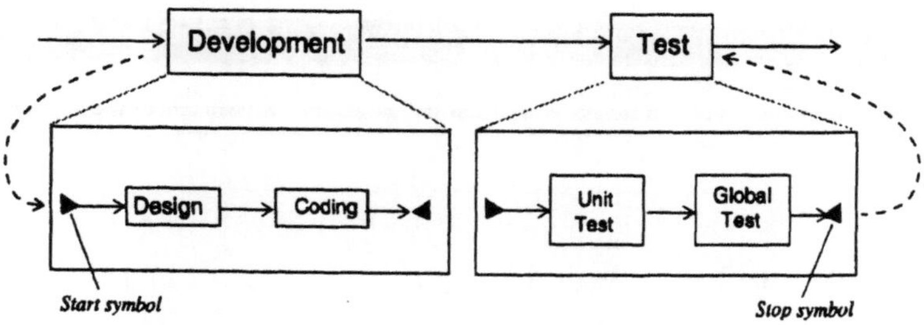

Figure 14. Activity decomposition.

The *Test* activity is composed of *Unit Test* and *Global Test* activities. At the first level, only CF between *Development* and *Test* is described, the CFs between their corresponding sub-activities are defined in separate diagrams at a lower level.

One important issue is that the decomposition technique must guarantee consistency across boundaries between levels with respect to both control and data flows.

Across a parent activity and its lower CF/DF description, the following rules are observed:

- the CF passes down/up through the start/end symbol, as shown with the "virtual" dashed arrow in figure 14. Note that there is one single start (stop) symbol per CF diagram.
- the DF is passed based on lexical bindings between internal and external inputs/outputs. Making the analogy with a procedure of most programming languages, a CF/DF diagram of an activity can be seen as the activity definition with *external IO* (vs. formal parameters of procedure definition), thus, an activity defined within a CF/DF diagram can be seen as an activity call with *internal IO* interconnected through data flow (vs. actual parameters of procedure call) (see figure 15).

2.9. *Evaluation of the basic model*

We designed and implemented, in early 1995, a high level graphical language (called APEL V2 (Dami et al., 1995)) that initiated a substantial part of the ideas presented above, but that was still underspecified at that time. The tool was tested for more than a year in different industrial sites with real-size projects.

The first users' impressions were that the system is more powerful than most of the current graphical process systems they knew, having a formal semantics, including an event-based control mechanism, offering three different perspectives for process modeling (SD, CF and DF), in addition to the static description of all the classes of entities (type level).

However, like other graphical process formalisms, it suffers from a number of drawbacks and limitations. It was still not powerful enough for real execution support. It was too deterministic, relying too much on automation, with weak semantics in a number of aspects, especially for the data flow control and cooperation support. In addition, the basic concepts of SD, CF and DF were not sufficiently integrated, in particular the connection between SD and CF/DF did not take into account the abstraction levels where they are defined. Finally,

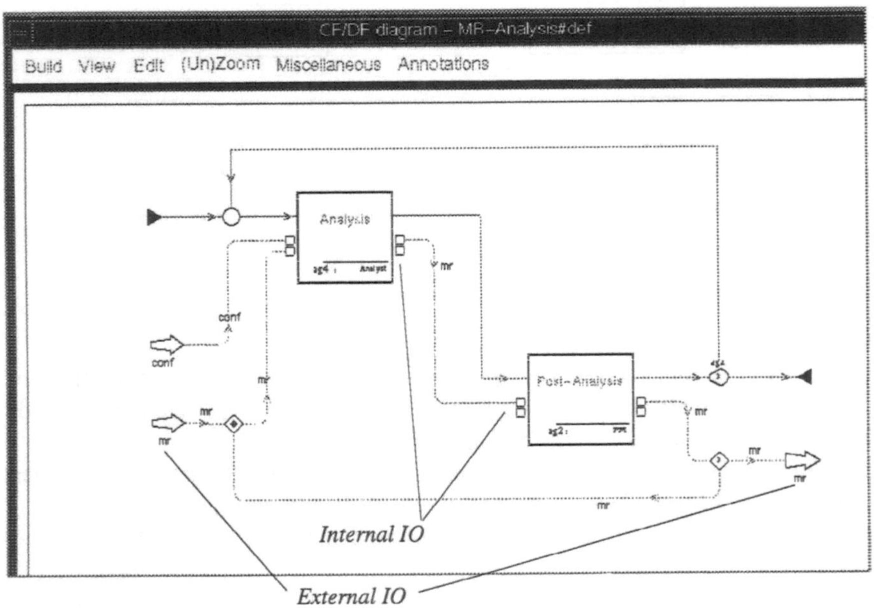

Figure 15. External vs. Internal IO.

no capabilities were provided at all for packaging and reuse, while support for human actors was deemed rudimentary.

Based on the feedback from our customers, the system was fully redesigned and extended in a number of ways. In the next section we discuss the advanced features that have been added to APEL, and the remaining issues that need to be resolved. For interested readers, a more detailed presentation of the current version, i.e., APEL v3, can be found in (Estublier and Dami, 1995).

3. The advanced features

3.1. The desktop paradigm

A practical process support environment must meet the main requirement that humans are the major actors in the process, and thus many decisions will (and must) remain under human control.

The *desktop* paradigm has been designed to meet that requirement. Each activity has a *desktop* on which the products managed by that activity, as well as its sub-activities[7] (including the control and data flows) can be manipulated.

Following the object oriented approach, methods can be executed on any object (be it a software artifact, a sub activity or an agent) present on the desktop, depending on their types, but the modelers can reduce the methods available in a given activity. For instance methods *create/read/delete/edit* are defined on the source code object type; they are all available in the *coding* activity but only the *read* method is available in the *test* activity.

In most traditional approaches like SADT (Marca and Macgowan, 1988), those activities that do not decompose are called terminal. Non terminal ones (called composites or complex) are purely abstract and have no real existence from a performance point of view the real work being executed at the terminal (leaf) ones exclusively. In our approach any activity, whether terminal or not, can be **active**, i.e., a human agent can execute methods on entities.

An activity description can be fully or partially defined. At one extreme, all data flows and control flows are explicitly defined; there is no need at all for human interaction from that activity point of view; it is a *fully automated* activity (e.g., a compile and build task). At the other extreme, all products and sub-activities appear on the desktop, but no data or control flow is defined; the human agent is in charge, using the available methods and tools, to perform its job freely; it is a *fully human-driven* activity (e.g., fixing a bug). Most often the reality lies somewhere between: a subset of the data and control flow is statically known and the rest consists of human decisions. Our system supports this, allowing any mixture of defined/undefined control and data flows.

It is important to note that this simple approach reconciles the major paradigms in an intuitive framework. It is the modelers' decision to define what is *automated* (explicit data and control flow) and what is *human driven* (calling methods); what is *deterministic* and what is *nondeterministic* (external event and users' method call); what is *prescriptive* (explicit control flow) and what is *proscriptive* (e.g., unavailable methods in the menus).

Figure 16 suggests that, at enaction time, opening the activity *A* opens the desktop of *A* and gives the agent *X* playing *role-A* access to *A* content, i.e., products, sub-activities (their associated CF/DF) and agents. From a practical point of view, it means that the agent *X* can apply the authorized methods on all these different entities (including *Change-State/Start/Stop/Suspend/Resume/Abort* of sub-activities if allowed).

In the example shown in the figure above, it is specified that as soon as the activity *A* starts, *A1* becomes active. However, *A3* can start only upon capture of event *Start*. The

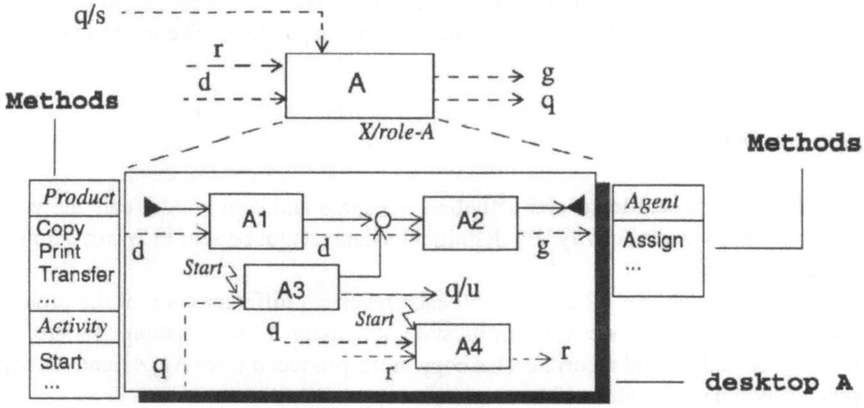

Figure 16. Activity and sub-activity desktop decompositions.

Start event is caused by an explicit method call (*Start* method) requested explicitly by the agent X. If the starting of *A3* is to be automated depending on the asynchronous reception of the data q, the event would have been (q, *input*, A); i.e., the event occurring upon the arrival of data q (*input* is the default state for an incoming asynchronous data).

Graphically, a data which is the origin or destination of a DF arrow (q and r) means it belongs to the current desktop. Thus it can be accessed and manipulated by agent X with respect to the authorized methods. In the example the data q, after being processed by *A3* is returned to the desktop of A, where agent X can, if required, apply the allowed methods (*edit*(q), *compile*(q), etc. for example) to it. Agent X can also decide to provide it as input to activity *A4*. Conversely, the data flow for *A1, A2*, being fully specified at model level cannot be changed by agent X and is fully automated.

3.2. Concurrent work

Any activity, be it atomic or composite, manipulates data that is maintained and managed into a desktop. The desktop is thus the data container of an activity.

We have assumed until now that a data can be processed by a single activity at a time; i.e., it is present on a single desktop at a time. In a number of cases, it is required that a data be processed concurrently by different activities, each one receiving a copy rather than the actual data as such.

In APEL, three modes for data circulation are possible. There are **Transfer**, **Share**, and **Copy**.

- **Transfer** means the data item is moved from the desktop of the source activity to the desktop of the destination activity. After the move, the data is no longer available to the source activity. It is the default mode denoted so far by: --->
- **Share** means the source and destination activities will access physically the same data. Concurrent changes may interfere in this case. The graphical notation is: -/->
- **Copy** means a copy of the data is sent from the source activity to the destination activities. In this case both the source and destination activities will have their own copies, so that both may perform (local) changes concurrently but separately. The graphical notation is: -//->

Multiple sending

When a data item is sent to several activities, the share and copy modes can apply independently for each target activity. By definition, a data cannot be sent in *Transfer* mode to more than one activity.

Figure 17(a) means that *A2* and *A3* will each receive a different copy of the product a (under names b and c respectively) in its state at activity *A1* termination. Figure 17(b) depicts that *A2* and *A3* will receive each a copy of the product a from *A1*; *A1* and *A3* copies are physically the same, while *A2* has a private copy.

In APEL, only a single copy of a data item is allowed in a desktop; i.e., if a second copy is placed in a desktop, both copies are automatically merged into a single one. This

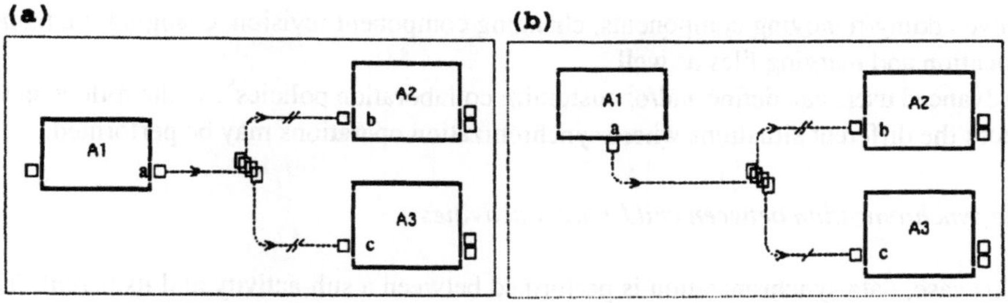

Figure 17. Split-up data flow.

is the case for example when a sub-activity returns a product to the parent activity which contains a copy of it. Note finally that the copies created in an activity are either deleted or resynchronized at activity termination.

3.3. *Collaboration policies*

Whenever multiple copies of a product are handled, there is a need to control the changes allowed on each copy. APEL provides standard policies which, apart from the input mode, describe more precisely how different copies are accessed in a concurrent way:

- **Exclusive.** When two activities own the same (logical) product both can, potentially, change it. The exclusive policy means that only one activity, whatever it is, can change it but not two or more. The graphical representation is: $\overset{x}{\dashrightarrow}$
- **Read only.** The data provided to the activity is read only, drawn: $\overset{r}{\dashrightarrow}$
- **Write.** The data provided to the activity is in read and write access. It is the default mode, drawn: ---->

Whenever concurrent changes are allowed, there might be a need for resynchronizing these copies. Synchronizing two copies means merging both of them, and replacing one copy by the result of the merge.

APEL provides a basic primitive allowing two copies *C1* and *C2* of the same product *X* owned by two activities *A* and *B* to be synchronized:

```
Synch(C1, A, B, C2)
```

which can be worded "synchronize product *C1* in *A* according to product *C2* in *B*". Formally, it is written:

```
C1(A) := C2(A) @ C1(B)
```

where "@" is the merge operator. It is important to note that *X* may be any product, including complex products (products that are composed of other products; for example a module which is composed of a header and a body). A merge does not only involve the classical file merge, but also any other attribute change, including composition relationship changes. For example, resynchronizing two copies of the same configuration may involve adding/removing components, changing component revision, changing file names of location and merging files as well.

Advanced users can define and/or customize collaboration policies[8]. In the following we discuss the different situations where synchronization operations may be performed.

Data synchronization between child-parent activities

In this case, data synchronization is performed between a sub-activity and its parent. The synchronization direction (child to parent, or the other way around) determines which of the sub-activity or parent desktop will be changed.

Another piece of information that has to be specified is the moment at which the data synchronization is to be performed. Typically, this can be determined by an event capture which governs a rule of the following form:

```
<event-capture> => <policy> ( <product> )
```

Where <policy> is a synchronization primitive name (e.g., *Promote, Resynch*, or any other user-defined policy), and <product> is the product which is to be synchronized.

The only mandatory synchronization policy is the merge of two copies of the same product when they "join" each other in the same desktop. Naturally, this happens between an activity and its sub-activity, at the sub-activity termination.

Graphically, a synchronization may be specified during a data transfer by drawing an arrow across the data flow. A child-to-parent synchronization is drawn by an up arrow: ---↑---

A parent-to-child synchronization is drawn as a down arrow ---↓--- . A bi-directional synchronization is also possible and is drawn as a bi-directional arrow.

In addition to data synchronization that can be specified in tandem with the data flow, it is also possible to force data synchronization during an activity execution. This is particularly useful for terminal activities whose data are not visible. In this case, the information is entered as a rule and attached to the activity input.

For example (figure 18), suppose the *Test* and *Coding* activities are executed in parallel. Each receives the configuration in input; the *Coding* activity fixes errors found in the *Test* activity and commented on in the error report (*ErrRep*) document. The policy says that as soon as a *Conf* reaches the state *valid*, it is promoted to the main activity, i.e., *Development*. On the *Test* activity side, the policy specifies that as soon as a new *Conf* is promoted to main activity, the local copy must be resynchronized with it.

Data synchronization between peer-to-peer activities

More often, collaboration can take place directly between sub-activities without going through the parent activity. In APEL any pair of activities holding copies of the same product can be directly synchronized. Such collaboration is drawn by a thick dashed arrow to which (as with the other collaboration types) the event and the collaboration primitive are attached.

Figure 19 shows how the synchronization of the previous example can be set up directly between the *Test* and *Coding* activities without going through the parent activity. Since the

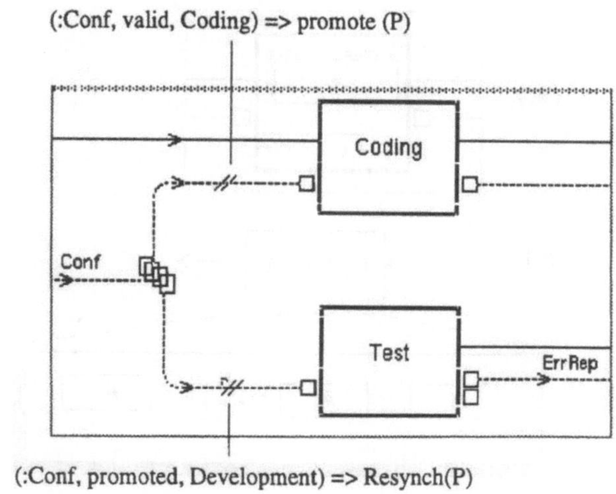

Figure 18. Asynchronous child-parent collaboration.

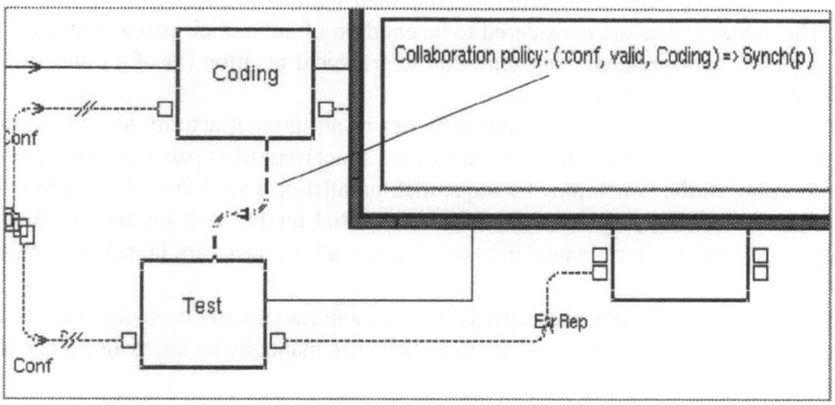

Figure 19. Peer-to-peer collaboration.

Conf copy is unchanged in the *Test* activity, the *Synch* operation simply replaces it by the copy of the *Coding* activity each time it is validated.

3.4. *Multi-instance activities*

Multi-instance activity is useful when the number of activity instances to be created at enaction time is not known in advance. This means that, contrary to a simple activity which is normally instantiated once at enaction time, a multiple activity can be instantiated a number of times.

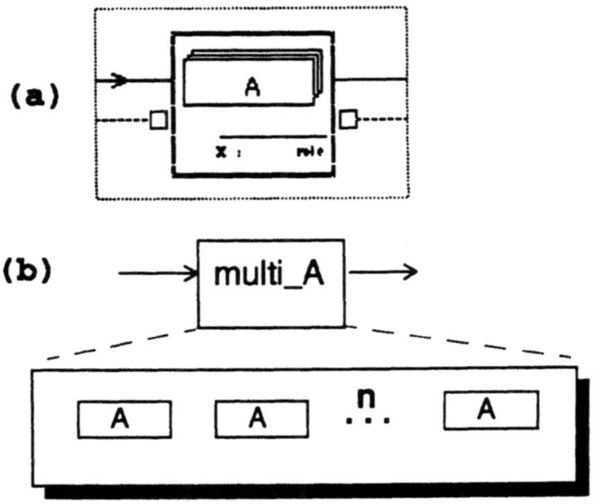

Figure 20. Multi-instance activity interpretation.

A multi-instance activity is treated as a composite activity, with the assumption that all its sub-activities are exactly the same, but that their number is undetermined at modeling time. The sub-activities are considered to be children of an implicit parent activity which is supposed to manage them. In figure 20 the graphical notation (a) of a multi-instance activity is interpreted as shown in (b).

The 'n' activity instances *A* are sub-activities of an implicit activity named automatically *multi_A*. The way the activity-instances are synchronized is partially determined by the **activation mode**, which may be sequential, parallel or asynchronous. Sequential and parallel modes are fully automated: in the sequential mode, each instance is launched when the previous one terminates; in parallel mode, all instances are launched simultaneously.

In the asynchronous mode, each activity instance will start upon occurrence of a particular event. If not specified, the instances are to be launched manually by the assigned agent ('x' in figure 20).

Example

Let us take the following scenario of a review (multi) activity. The review activity receives documents through asynchronous data flow. These documents have to be reviewed on the basis of common requirements. Each time a new document is received (in state to*Review*), a review activity is launched with the following inputs:

- the requirement document (*req*): It is to be read only. Synchronous input is required with *shared* mode (no synchronization).
- the document to review (*doc*): It is received in read only access and in an asynchronous way with *Transfer* mode.

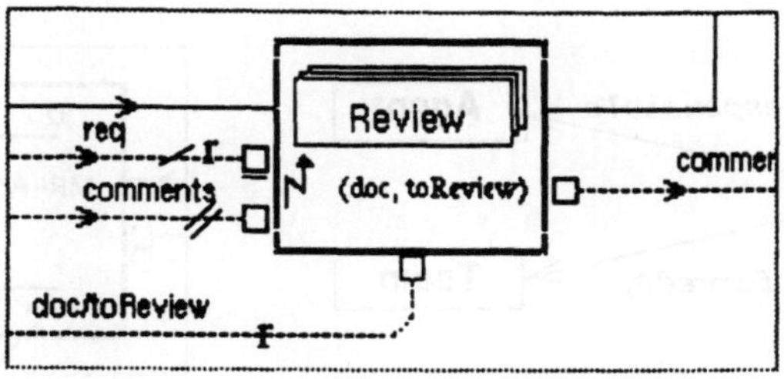

Figure 21. A simple review scenario.

- the *comments* document (initially empty): It is to be filled by each reviewer. Again, synchronous input is required but with *Copy* mode.

This scenario can be represented in APEL as shown in figure 21.[9]

3.5. *Organization and agent modeling*

For large software projects, performers may be organized into teams. A team can itself be broken down into sub-teams and so on. For example, teams for a software project may be organized into a *Development* team, *Maintenance* team and so on. The Development team itself may include a *Coding* team and a *Test* team. Such human resource organization is praised by the Actor Dependency model (Yu and Mylopoulos, 1993, 1994; Brian et al., 1988).

In APEL, the human resource organization for a process is modeled in terms of teams, positions, roles and agents as well as relationships between these elements (see figure 22).

A **Team** consists of positions that are required for the team tasks. It may be composed of other (sub) teams.

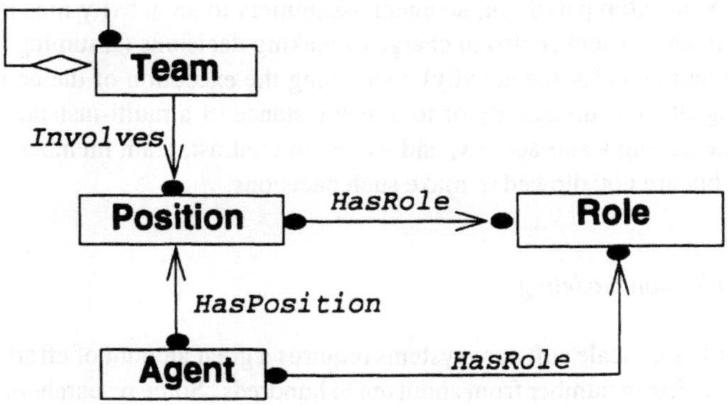

Figure 22. Team and agent meta model.

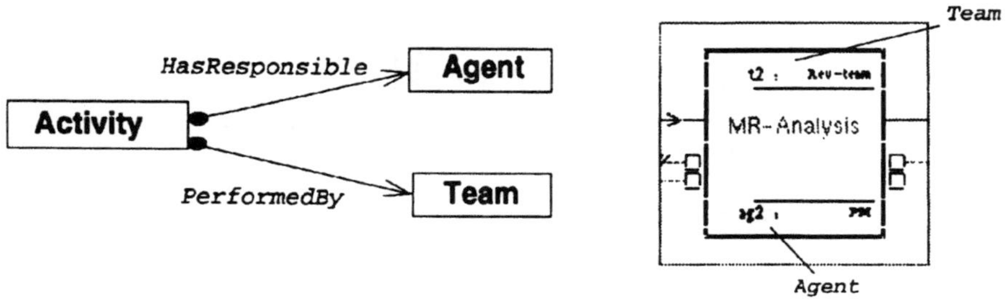

Figure 23. Relating process and organization models.

A **Position** is a place holder for a person in the organization structure. Examples of positions are *Project Manager*, *Software Engineer*, *Quality Expert*, etc. One position can have several roles.

A **Role** defines the function/responsibility of a human agent in the realization of a specific task. Performers holding the same position (e.g., *Software Engineers*) may play different roles in the same process (e.g., coder and tester).

Agents are human individuals who are actually performing activities during the process execution. An agent has a single position in an organization, but may play different roles when involved in a project.

For all of the concepts described above the user may define attributes as for product types. Typically, these attributes can then be used for role constraints. For example, one may define the attribute *Qualification* that allows the skill areas of an agent to be described, and define a role which requires a particular qualification that an agent must have to fit.

In addition, APEL relates the process model with the organizational context in which it will be enacted. This is done by establishing relationships between the entities of the process model (activities, products) and the entities of the organizational model (agent and team types), as presented in figure 23. This is used at enaction time to assign teams and individual agents to activities by instanciating the specified agent types or team types.

Following the desktop paradigm, an agent assignment to an activity means the agent is involved in that activity and is also in charge of making decisions (assuming the agent has appropriate permissions for the activity) concerning the execution of the activity, such as assigning an agent to a sub-activity or to a new instance of a multi-instance activity (see Section 3.4), launching a sub-activity, and so on. In contrast, team members are involved in the activity but are not allowed to make such decisions.

3.6. *Communication modeling*

Development of large scale software systems requires a great amount of effort and involves many people ranging in number from about ten to hundreds. Some researchers estimate that developers spend 70% of their time working with each other (Demarco and Lister, 1987), and that activities conducted by teams represent 70% of the cost of large software systems (Jones, 1986).

In APEL we refer to the work of teams as *cooperation*. Cooperation has three aspects, some of which have already been discussed: *Coordination* is related to activities ordering and synchronisation; *Collaboration* is related to the control of information and data sharing; and *Communication* is the explicit exchange of information between activities/agents.

In APEL *coordination* is modeled through control flow (Section 2.5) which describes the sequencing and the synchronization of activities that make up the process. *Collaboration* is modeled through data flow (Sections 2.6 and 3.3) which describes how to deal with the sequential or simultaneous accesses to the same set of products. *Communication* is intended to meet the following requirements:

- agents and activities in the process must be able to ask each other for services,
- they must be able to exchange information explicitly and whenever they want,
- they must be notified, if required, about the state of the process.

To some extent, communication could be implemented through data flow. The idea behind an explicit communication modeling is to provide easy and flexible ways to express at conceptual level how activities/agents (should) communicate with each other, in order to suggest/impose a communication policy at the execution level (which messages are allowed, how to react when receiving a message, etc.).

We model communication as the exchange of "short-lived" *messages* between activities/agents, while data flow is the exchange of "long-lasting" *products* between activities. Data flow is deterministic; it means that in each data transfer, the source and destination must be statically known; for communication, the destinations of a message are controlled by dynamic *subscriptions*.

A message is an entity that appears as a form. Messages are exclusively handled through a form manager. They are sent by specific primitives (e.g., *Send*) to be received in agent or activities *Agendas*. An agenda is associated with each agent and with each activity. It may contain zero, or more activities or messages at a given point in time. By agenda we mean the place (in practice a window) where an agent or an activity finds the (sub)activities it has to perform, as well as messages sent to them. It is worth noting that, depending on whether the receiver is an agent or an activity, the message, when received, will be visible either directly in the agent's agenda or in the activity in which it is supposed to reside.

Two kinds of messages are distinguished: *Notifications* and *Requests*. A notification is a means by which activities/agents communicate/inform each other without requiring an answer from the receivers. For example, an agent in an activity may send a notification to inform that he has terminated his activity. A request allows agents/activities to ask for each other's services. A service may be providing information, executing a task, assigning an agent to an activity, etc. In contrast to notifications, a request requires answers from receivers. To answer a request, three policies may be used. The first requires all receivers to handle the request by responding to it. The second policy requires a specific minimum number of responses without specifying from whom. The last policy consists of naming explicitly the receivers who must answer the request. When a request is handled, the reply is sent directly to the sender as a notification message.

Figure 24 illustrates the principle of communication in APEL. A message can be sent to an explicitly named receiver (either an agent or an activity), in which case only the agenda

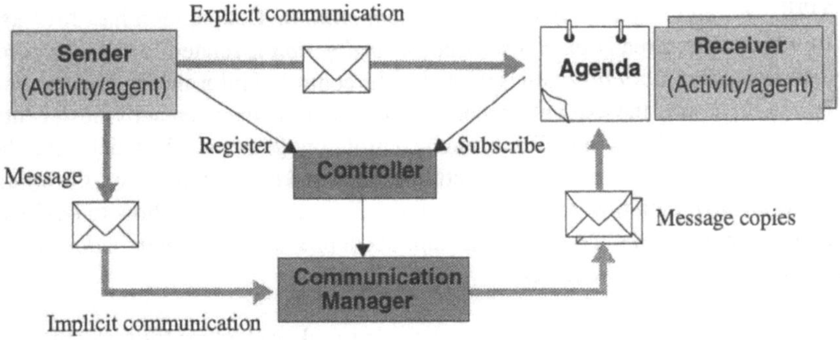

Figure 24. Communication fundamentals.

of that receiver will get the message. This form of communication may turn out to be too limited and constraining, particularly in those situations where sending a message must be anonymous, dynamically reconfigurable, to multiple receivers, and so on. APEL makes this possible by providing a message **subscription** mechanism in addition to explicit sending. This way a message can be disseminated with no obligation for the sender to know the receiver(s). Agents (or activities) who are interested in a message must subscribe first so that they can "capture" it.

Subscription can be expressed in terms of a message type, a name of a particular message or a message content (the values of certain message attributes). In this approach, widely used in message servers (such as Field (Reiss, 1990), Tooltalk (Frankel, 1990), Corba (Otte et al., 1996)), sender and receivers do not know each other directly. Dynamic subscription is supported by allowing the addition or removal of subscriptions during process execution.

Messages are represented graphically by envelopes. A message exchange is represented by a dashed line from the sender to the receiver(s). To distinguish between a message exchange and a DF, an envelop is indicated on the dashed arrow representing the message exchange. A subscription is represented by an expression of the form *<[message_name]: message_type [(pattern)]>* where the first and the third terms of the expression are optional.

Figure 25 illustrates how communication is modeled in APEL. In the example, error reports (of type *ErrorReport*) routing is statically and explicitly drawn between *test* and *dev* activities. The *dev* activity can receive error reports from any *test* activity, and then sends back comments (of type *Comments*) without designating the receivers. In order to actually receive only those comments related with the error reports it has sent, each *test* activity subscribes to: *Comments (Self.ERnumber IN AllMyER)*. *Self* refers to the comment being

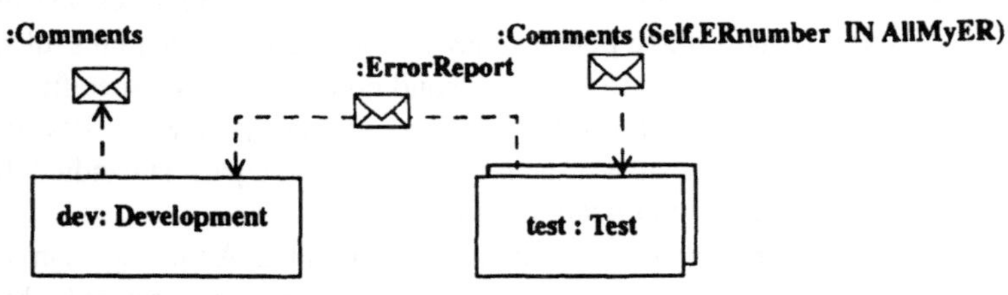

Figure 25. Subscription and message passing.

sent. *ERnumber* is a field of the comment containing the name (number) of the error report to which it corresponds. *AllMyER* represents the set of all error reports sent by that *test* activity.

3.7. Measures and versions

It is widely agreed that process improvement must be a fundamental and strategic goal for any company. Such an improvement can only be achieved based on quantitative and objective information. Up to now the only available objective information was an external assessment of the current company process maturity (CMM). While such evaluation is clearly useful, we also believe in systematic process improvement based on goal-oriented measurement, monitoring, evaluation, packaging and reuse.

An improvement program involves the definition of quantitative quality goals by refining these goals into metrics in a top-down fashion. Metrics are integrated into products and processes as well, so that during the process execution they can be collected, and then analyzed in order to learn how to better proceed for future projects. In APEL, the GQM (Goal Question Metric) approach from Basili et al. (1994) (Rombach, 1991) was chosen. GQM is an approach for goal-oriented measurement in software projects which supports measurement of products and processes. The principles upon which the GQM approach is built include:

- Set explicit *measurement-goals* by considering the *object* (e.g., product or process) under study, the *purpose* (why) behind the measurement goal, the *focus* (what) (e.g., reliability, effort), the *viewpoint* (e.g., manager, customer), and the *environment* that determines the context of the measurement program (e.g., project P, or division D).
- Acquire *quality models* from people involved, based on interviews, using "GQM abstraction sheets".
- Derive appropriate *metrics* (through questions). The GQM plan consists of a goal (G), questions (Q) which operationalize the goal, and metrics (M) which quantify the questions. See an example in figure 26.
- Analyze data according to goals, through a feedback session, usually performed as brainstorming and discussion meetings.

These principles distinguish the GQM approach from other measurement approaches which do not sufficiently consider the importance of explicitly identifying goals and the viewpoints of the people involved in the software project of interest. As a consequence, measurement may not be focused enough and the personnel may not be motivated to provide the measurement data.

APEL provides support for:

- building GQM models; that is, defining goals and deriving metrics for the related process. Figure 26 shows a snapshot fragment of a GQM model including the different elements involved in the GQM approach, such as goal, abstract sheets, questions and metrics.

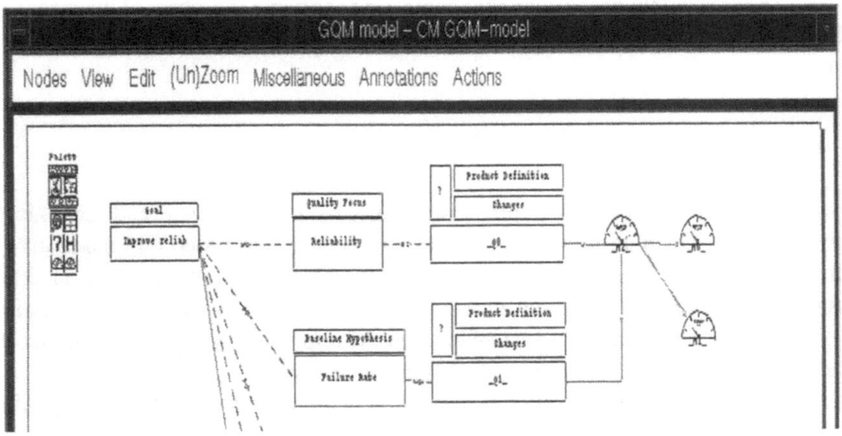

Figure 26. A GQM plan (or model).

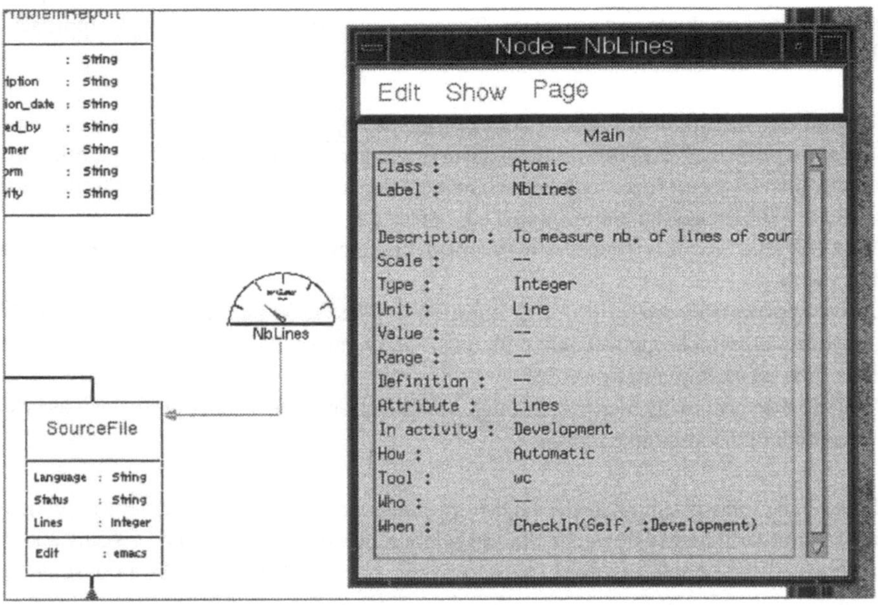

Figure 27. An example of metric.

- integrating the defined metrics into products, activities and agents. In addition, the APEL GQM editor allows modelers to define metrics aggregations in which complex metrics can be defined based on atomic ones. An example of a metric definition and its integration into a product type is depicted in figure 27.

The definition of a metric gives also a concise specification of the data collection process. For each metric of the GQM plan, it is specified:

- what data is collected,
  In the example of figure 27, attribute *Lines* of any product of type *SourceFile*.
- when a particular piece of data should be collected,
  E.g., (the When attribute), each time a source code is checked in during the development activity (*checkIn(Self, Development)*).
- how data is collected,
  Data can be collected automatically using measurement tools (in the example *wc* unix command), or manually using data collection forms which can be presented on-line on a computer screen or as printed questionnaires.
- and by whom.
  In case the data is collected automatically.

3.8. Concept integration and reuse

APEL supports both product-centered (via SD) and activity-centered (via activities CF/DF) approaches. Depending on the domain, background, and level of abstraction the two approaches have strengths and weaknesses in such a way that they become complementary. The product-centered approach is intuitive and satisfactory when product control is a major concern. Not surprisingly, SD are by far the most used technique in the software configuration management field, where the experience has shown that modelers think about their processes in terms of product evolution states.

However, this approach has proved useful only at a rather low level of abstraction. This weakness makes this approach inappropriate for capture because it does not supply the user with a global view of the process. On the other hand, the activity-centered approach seems appropriate for capture: it matches programming habits, and can describe processes at different levels of abstraction. Moreover, it is convenient for business applications where the focus is on the workflow.

From a reuse point of view, the integration of both concepts gives rise to the crucial question of how to consider them in a consistent way knowing that they are cross-dependent.

In APEL, an activity is viewed as a black box receiving artifacts as input and transforming/producing artifacts as output. This definition is intuitive for users and clearly shows an *activity* to be a *procedure*. Within such an approach the best way to promote reuse is to borrow the concept of encapsulation, which consists of defining an activity as a black box by hiding its *body* (i.e., its internal implementation), and "exporting" its *interface* (i.e., its signature) so that the activity can be used from the outside. In APEL an activity interface is defined through the activity inputs and outputs (the signature), and the external events that the activity produces or has to react towards. Adopting the concept of encapsulation means that artifacts and their states may have different names, depending whether they are referred to inside or from outside the activity (formal vs. effective parameters).

On the other hand, classically, an SD defines the product evolution in a global way: states are global and transitions are defined independently from the activities. The behavior of an activity, with respect to its local data, is visible from any other activity. The encapsulation principle governing activity reuse is broken.

Integrating state diagram with activities

The main features of Harel's diagrams from which the APEL's state diagram (SD) is inspired, are the concept of depth (OR decomposition) and orthogonality (AND decomposition). These concepts allowed us to deal with two crucial concerns: how to depict the behavior of products at different levels of abstraction and how to represent their concurrent evolution.

Keeping track of the activity decomposition process, we have introduced the notion of *local SD*. A local SD defines the state changes of a given product type *within a particular activity*. The local states can be defined without any assumption about where the activity will be used. Thus state decomposition follows activities' decomposition exactly. However, contrary to a non-terminal activity, a composed state is purely abstract in the sense that it does not indicate the actual state of the product, but instead just indicates that the product is currently handled in a specific activity. This is why an abstract state is always automatically named by concatenating the name of the activity it is attached to with the prefix "In_".

Figure 28 shows an SD defined for a product named *Conf.* At the highest level (*Develop* activity) the product may be in states baseline, *In_Code*, *In_Test* and *integrated. In_Code* and *In_Test* are abstract states which means when the product *Conf* is in one of the two states, then it is currently handled in one of the two corresponding activities, i.e., *Code* or *Test* respectively. At a lower level, the two abstract states *In_code* and *In_test* are refined into sub-SDs which describe more specifically how the product *Conf* evolves within the two activities *Code* and *Test* (through states *coded, compiled*, and so on).

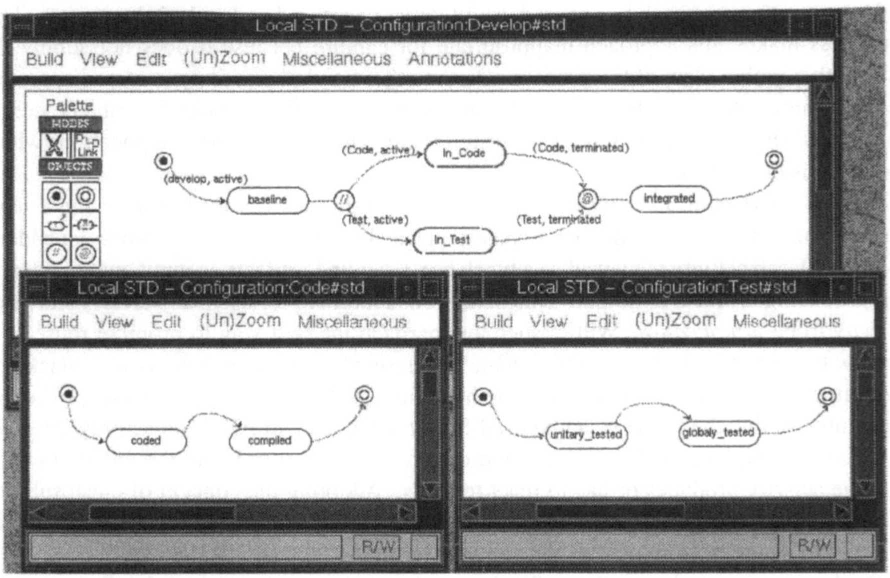

Figure 28. Sub-SDs and "concurrent" SDs.

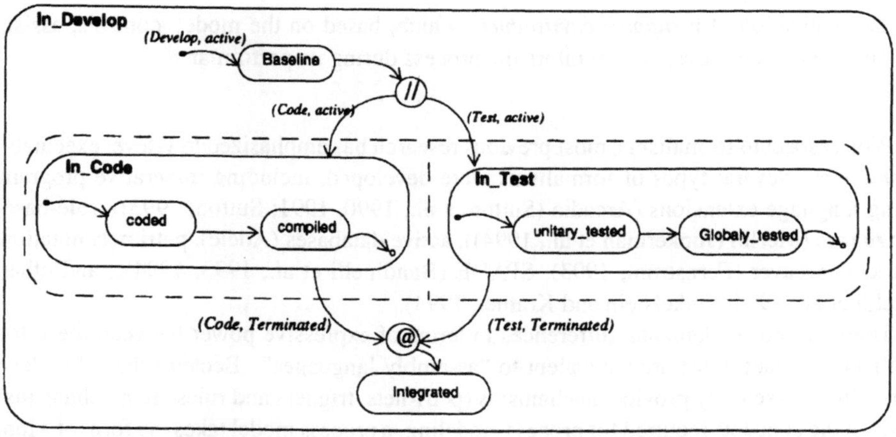

Figure 29. Concurrent states and local SDs.

The same *Conf* may have different and concurrent behaviors when different copies are handled simultaneously in *Develop*, *Code* and *Test* activities. Each copy evolves independently from the others, i.e., each copy follows an independent local SD. In order to model these parallel evolutions we have adopted the concept of orthogonality which we have extended to represent synchronization of copies. The symbol "@" represents the synchronization operation; the symbol "//" indicates that the following sub-SDs hold for different copies evolving in parallel.

The entirely expanded SD can be generated automatically and displayed as shown in the figure 29.

The dashed boxes contain the concurrent states. Inside, *Conf* is simultaneously in *In_Code* and *In_Test* states. Leaving the dashed area forces a resynchronization of the two copies.

The fact that a product type can have a structured/modularized SD per activity allows concise reuse of process fragments relying on their products' evolution. For instance, integrating a new activity including its product SD into a process does not have side effects on the other parts. Another advantage is that the capture of an SD for a product may be considered with respect to a particular point of view and at a given level of abstraction.

4. Related work

Software process modeling and support have been an active research area over the last decade (and continues to be so). However, it is not yet a mature field because there is as yet no consensus on terminology, problems and solutions, and many important concerns have not been resolved in a satisfactory way.

Globally, the problems to solve are:

1. the definition of *formalism*(s) in which it is possible to describe (model), with sufficient detail, how a (software or other) product must (should) be produced, and

2. the realization of a *support environment* which, based on the model, controls, assists, automates, measures, and monitors the process during its performance.

With respect to formalisms, most previous research has emphasized low-level executable *formalism*. Several types of formalisms were developed, including imperative programming language extensions (Arcadia (Sutton et al., 1990, 1991; Sutton, 1993)), rule-based languages (Merlin (Junkerman et al., 1994)), active databases (Adele), petri nets notations (process weaver (Fernstrom, 1993), SPADE (Bandinelli et al., 1993, 1994)), and others (Baldi et al., 1994; Finkelstein and Kramer, 1994).

There are no fundamental differences in terms of expressive power between these formalisms. In fact, most are equivalent to "assembly languages". Because these low-level formalisms essentialy provide mechanisms (petri-nets, triggers and rules are mechanisms), and not the concepts required for process modeling, a process model takes the form of a long and confusing program from which it is difficult to get an overall and clear picture of the process. This defeats the original purpose of process modeling, which is to provide a clear and unambiguous description of the process, to be able to evaluate and compare models, to disseminate, to understand and to improve a process. None of these objectives can be achieved with a low-level formalism. Clearly there is a need for high level formalisms.

The notations provided by many workflow management systems tend to be high-level and graphical, and the user interfaces they provide tend to be user friendly. This accounts for their success (Georgakopoulos et al., 1995; Gruhn, 1994; Mohan et al., 1995; Bolcer and Taylor, 1996). A few software process support systems followed this approach by providing high level formalisms. Unfortunately, high-level formalisms, and notably graphical ones, are often semantics-free and thus ambiguous and non executable which substantially reduces their usefulness (Barghouti et al., 1995).

APEL bridges the gap between semantics-free high level formalisms and low level executable ones. We believe that providing such a broad scope language is a mandatory step toward a widely accepted software process support environment.

The software process aspects generally tackled are related to very low-level tasks, edit/compile being a classic example. We do not believe engineers need to be supported in their usual individual activities; they know their job pretty well and automating compilations or modeling in detail how to find a bug can only be counter-productive. In fact, the real difficulties come from cooperative work and organizational aspects which are ill supported by current formalisms, while they account for up to 70% of the effort (Demarco and Lister, 1987).

This aspect has been addressed by work in CSCW (Malone et al., 1992; Roseman and Greenberg, 1992; Conklin and Begeman, 1988; Paoli and Tisato, 1994; Grudin, 1994; Reinhard et al., 1994), although in a synchronous way. Our work differs in that collaboration is essentially asynchronous; concurrent work (like debugging) on the same objects is not performed by concurrent editing (in the program source code), as in CSCW, instead, APEL provides each activity with an isolated work space (a different desktop) and, later on, resynchronize. This is why a large part of our work addresses the management and synchronization of concurrent isolated work.

In Oz (Ben-Shaul and Kaiser, 1994) cooperation is addressed in another way. The issue is making distributed environments interoperate using the paradigm of the international alliance (Ben-Shaul and Kaiser, 1996). Oz allows flexible collaboration among independent environments in the "share nothing" mode, rather than defining and supporting the collaboration among people in the same team. We have not addressed the distribution and multiple environment interoperability issues in our current prototype, but it is now an active research objective in our team.

It was discovered early on that real software processes are large and complex. This is why there is a need to provide structuring concepts and techniques for modularization purposes. Work on the view concept (Estublier and Belkhatir, 1995; Verlage, 1994; Finkelstein et al., 1993; Nuseibeh and Finkelstein, 1992) as well as on the definition of clear process interfaces (Avrilionis et al., 1996), are moves in this direction. But currently this work is very much in a preliminary stage. Scalability constitutes a real challenge for software process tools. We think the proposed architecture (not detailed here) is both a viable and scalable solution for multi-view support.

In this work, we did real efforts to promote reuse. The interface definition, the encapsulation (of data, states and sub activities), the integration of data and activity aspects are steps in this direction (Estublier and Dami, 1996a). However our experience shows that the time for reuse has not arrived yet. Users currently develop model instances (i.e., that fit a given project process), and then generalize (to try to fit a range of similar projects), rather than the contrary. More work and experience is needed before a convenient solution can be found.

What makes the software process a specific domain is that it deals with humans and thus with uncertainty. The non-deterministic behavior of humans, their creative nature in designing and building software, the long duration of the tasks, along with the unavoidable changes during execution, are the core difficulties. Most tools either represent human as tools (!), or do not model them at all. Most often human support is limited to agenda management; which is needed but insufficient. These topics are being actively investigated, but currently solutions are most often not satisfactory.

We believe that the desktop paradigm proposed here is a good solution, being intuitive and allowing undefined processes (fully human-driven), under-defined processes (partly human-driven) and completely defined processes (fully automated) to be combined freely.

5. Conclusion

The goal of this work was not to explore new issues in software process modeling, but to define and develop a practical high-level and broad-scope process formalism. This goal is ambitious and opens up a number of issues. We had to investigate how different perspectives can be managed simultaneously and consistently. More critically, we had to find new solutions for integrating different paradigms, which are sometimes hard to reconcile, in an intuitive way (prescriptive/proscriptive, deterministic/non-deterministic, automated/human-driven). Taking advantage of our original expertise in software configuration management, we also had to face the problem of how to address issues about the management of multiple copies and collaboration policies, whilst still offering high level capabilities that facilitate work of process modelers.

Implementation and current state

The APEL editor is implemented using the GraphTalk[10] meta editor. It was released in April '96 and has been used since, for evaluation purposes, in a number of companies including Bosch, Siemens, Q-Lab, and Dassault Systems. It is expected to have industrial use in real projets.

The APEL compiler works as follows. First, the APEL graphical representations are translated into the APEL textual representation by means of GQL (Graphtalk Query Language) requests. This APEL representation is then compiled into the Adele language. The type level is directly translated into an Adele data schema; APEL rules into Adele triggers; methods into Adele methods. The multiple copies control, the merges and the collaboration policies are supported by the Adele work-space manager (Estublier, 1996). The first version of the compiler has been running since June '96; the second version, released September '96, generates Adele executable code for almost all the APEL features presented here. In other words, once modeled through the APEL graphical editor, there is no need to write a single line of code to fully execute the model.

The desktop, including message-based communication and agenda manager as well as the monitoring tool were developed in Java and have been available since June '96. A complete APEL architecture has been built including secondary models for instantiation, user interface, monitoring, and error management (see figure 1). All these secondary models and related tools have been developed in Java. The complete environment has been under experimentation since October '96.

Debugging, monitoring and measures are current research and implementation hot topics, as well as the definition of a very general and flexible architecture supporting large scale distribution and heterogeneous federation of sub environments and formalisms.

APEL has the ambitious goal of being "wide scope" and "platform independent". These goals are pretty close to become reality. The concepts of APEL have been evaluated in fields other than software production, and a number of implementations are planned, applying the same editor (i.e., the same meta model) to model different kind of processes, and compiling the resulting model toward different enation support tools.

In most of these experiments, only the pertinent part of APEL is supported. For example, the compiler between APEL and Process Weaver, for workflow support, does not support data flow and desktop synchronization. Manufacturing processes (the process of building complex artifacts like a car engine) have been described in APEL, and their simulation will be suported by compiling the model into Esterel. Here too, a number of APEL features are not implemented, most notably everything related with non determinism and human driven processes. Another experiment, in the domain of virtual product definition processes is under way. These processes describe concurent engineering when designing complex artifacts. The corresponding process support would be provided by different CAD components.

Open issues and future work

A number of open issues are still under investigation.

First, the capture facilities were not high on our priority list, and thus they were less developed than the other capabilities. The reason is that it was unclear for a long time how to enact a high level, flexible formalism. However, customer feedback shows that this aspect is of primary importance for user acceptance. Thus, we are currently exploring more graphical flexibility, more informal modeling, and a way to transform gradually an informal model into a formal one. In particular, we are investigating the integration of Improvise (Barghouti et al., 1995), which provides visualization and multimedia facilities, with APEL.

Reuse is another topic of primary importance which has still to be improved. We expect to define a (meta) environment for the support of packaging and reuse (Estublier, 1992; Belkhatir and Estublier, 1996; Estublier and Dami, 1996a), better definition of the interface concept and component substitution capability.

The approach based on multiple points of view is currently being investigated (Estublier and Belkhatir, 1995; Promoter Group, 1996). Owing to the fact that the work on multi-view is still in its infancy, the results have still not been integrated. However, we can say that the notion of secondary models implements a kind of multi-viewpoint approach.

On the practical side, the design of the APEL environment gave rise to numerous difficulties, particularly that of providing models which represent the problem to solve at high level (this is what capture does), and programs which represent a solution to the problem (this is what execution is). The architecture which clearly separates different models and formalisms, one being the main model, and the rest implementing different aspects of the technical solution, seems very promising in this respect. The current work includes the extensions of APEL toward an open, distributed and heterogeneous environment.

Throughout the APEL project, we were very careful in taking design decisions which will not hamper, in the future, the scientific development of the work. It seems this was also successful, and the environment proved to be very useful for designing, developing and experimenting with challenging prototypes.

We have realized an APEL environment dedicated to software processes, and a number of other APEL environments are about to start for other domains. This family of APEL execution environments are sharing the APEL concepts and editor, but each one is providing a different compiler for a different support tool. Research work on architecture and interoperability is under-way to address in a more general way how these environments could share more than the bare editor, and how these environments could collaborate to provide an end to end support environment.

Acknowledgments

We are grateful to Hervé Cimadomo who realized the run time support; Pascal Palazon, Stephane Valette, Nicolas Garnier and Paulo Dacosta who realized the APEL compiler; Olivier Rufin and Stephane Berrin who realized the APEL agenda manager; to Maryse Lebrasseur, Philippe Pouzet and Jean Routin from Cap Gemini Innovation who contributed to the design and development of APEL V2, and to Naser Barghouti who did a close review of this paper and fixed our rough English.

Notes

1. PERFECT stands for Process Enhancement for Reduction of software deFECT. It is an ESPRIT project funded by EEC.

2. Almost all pictures are screen from APEL editor. A screen dump is always placed within a frame (greyed line), with sometimes additional comments written in italics.
3. An activity can be viewed as the abstraction of a set of concrete operations that are performed on objects through methods invocation.
4. Built-in methods are predefined methods needed for supporting the dynamic aspects of the APEL formalism (e.g., *Start*, *Stop*, *Transfer*, *ChangeState*, etc.)
5. Because they are predefined, activities SDs are always inherited as is and cannot be refined.
6. The desktop is the container of the objects manipulated by an activity. The concept of desktop is presented in Section 3.1.
7. In our model activities are themselves objects, with methods (*Start*, *Stop*, *Suspend*, *Abort*, *Resume* and *ChangeState*), attributes and relationships. Sub-activities are thus managed in the same way as the other entities present on the activity desktop.
8. In the standard library of policies, one can find automatic Resynch and Promote based on events, Resynch-Promote, Notifications about state changes, NotifyFreeze which keeps the current state for later manually requested resynchronization, and so on.
9. The cooresponding textual code is the following:

```
(doc, toReview, multi_Review); State(multi_Review) = active =>
        R := New(Review); Transfer(doc, multi_Review, R); C := New(Comment);
        Transfer(C, multi_Review, R); Start(R);
```

10. GraphTalk is a registered trademark of Parallax.

References

Avrilionis, D., Belkhatir, N., and Cunin, P.Y. 1996. A unified framework for software process enactment and improvement. *4th International Conference on the Software Process*, Brighton, England.

Baldi, M., Gai, S., Jaccheri, M.L., and Lago, P. 1994. E3: Object-oriented software process model. In A. Finkelstein and J. Kramer (Eds.), *Software Process Modelling and Technology*, John Willey and Sons, vol. 1.

Bandinelli, S., Baresi, L., Fuggetta, A., and Lavazza, L. 1993. Requirements and early experiences in the implementation of the SPADE repository. In W. Schaefer (Ed.), *Proc. of the 8th Int. Software Process Workshop*, Germany: IEEE Computer Society Press.

Bandinelli, S., Barga, M., Fuggetta, A., Ghezzi, C., and Lavazza, L. 1994. SPADE: An environment for software process analysis, design and enactment. In A. Finkelstein and J. Kramer (Eds.), *Software Process Modelling and Technology*, John Willey and Sons, vol. 1.

Barghouti, N.S., Koutsofios, E., and Cohen, E. 1995. Improvise: Interactive multimedia process visualisation environment. *Proceedings of the 5th European Software Engineering Conference*, Barcelona, Spain.

Basili, V.R., Caldiera, G., and Rombach, H.D. 1994. Goal-question-metric paradigm. *Encyclopedia of Software Engineering*, John Willey & Sons, vol. 1, pp. 469–476.

Belkhatir, N., Estublier, J., and Melo, W.L. 1993. Software process model and work space control in the Adele system. *Proc. of the 2nd Int. Conf. on the Software Process*, Berlin, Germany.

Belkhatir, N. and Estublier, J. 1996. Supporting reuse and configuration in large scale software process models. *10th International Software Process Workshop*, Dijon, France.

Ben-Shaul, I.Z. and Kaiser, G.E. 1994. A paradigm for decentralized process modelling and its realization in the Oz environment. *16th International Conference on Software Engineering*, Sorrento, Italy.

Ben-Shaul, I.Z. and Kaiser, G.E. 1996. Architecture for federation of process-centered environments. Technical report, CUCS-015-96, Columbia University, Department of Computer Science.

Bolcer, G.A. and Taylor, R.N. 1996. Endeavors: A process system integration infrastructure. *4th International Conference on the Software Process*, Brighton, England.

Booch, G. 1994. *Object Oriented Analysis and Design*. Benjamin-Cummings.

Brian, L. Melo, W., Seaman, C., and Basili, V. 1988. Characterizing and assessing a large scale software maintenance organization. *Proc. of the 17th Int. Conf. on Software Engineering*, Seattle, Washington, USA, pp. 133–143.

Coad, P. and Yourdon, E. 1991a. *Object Oriented Analysis* (Second edition), Yourdon Press, Prentice Hall.

Coad, P. and Yourdon, E. 1991b. *Object Oriented Design*. Yourdon Press, Prentice Hall.

Conklin, J. and Begeman, M. 1988. gIBIS: A hypertext tool for exploratory policy discussion. *Proc. of the ACM Conference on CSCW*, Portland, OR.

Dami, S., Pouzet, P., and Routin, J. 1995. Internal specification of APEL v2. Esprit deliverable, PERFECT project, LGI/IMAG, Grenoble, France.

Demarco, T. and Lister, T. 1987. *People-Ware*, New York: Dorset House.

Estublier, J. 1992. Reusing software processes. In J.-C. Derniame (Ed.), *Proc. of European Wokshop on Software Process Technology EWSPT2*, LNCS, Trondheim, Norway: Springer-Verlag, vol. 635.

Estublier, J. 1996. Work space management in software engineering environments. In I. Sommerville (Ed.), *Proc. of 6th Int. Workshop on Software Configuration Management*, LNCS, Berlin, Germany: Springer Verlag.

Estublier, J. and Casallas, R. 1994. *The Adele Software Configuration Manager*, ch. 4, pp. 99–139. *Trends in Software*. J. Wiley and Sons, Baffins Lane, Chichester West Sussex, PO19 1UD, England.

Estublier, J. and Belkhatir, N. 1995. A generalized multi view approach. In W. Shaeffer (Ed.), *European Workshop on Software Process Technology (EWSPT4)*, Leiden, The Netherlands: Springer Verlag, vol. LNCS 913, pp. 179–185.

Estublier, J. and Dami, S. 1995. APEL v3 specification. Esprit deliverable, PERFECT project, LSR/IMAG, Grenoble, France.

Estublier, J. and Dami, S. 1996a. About reuse in multi-paradigm process modelling approach. *10th International Software Process Workshop*, Dijon, France.

Estublier, J. and Dami, S. 1996b. Process engine interoperability: An experiment. In C. Montangero (Ed.), *European Workshop on Software Process Technology (EWSPT5)*, LNCS, Nancy, France: Springer Verlag.

Fernstrom, C. 1993. Process weaver: Adding process support to Unix. *Proc. of the 2nd Int. Conf. on the Software Process*, Berlin, Germany.

Finkelstein, A., Gabbay, D., Hunter, A., Kramer, J., and Nuseibeh, B. 1993. Inconsistency handling in multi perspective specification. In *Proc. ESEC 93*, Garmish, Germany, vol. LNCS 717, pp. 84–99 .

Finkelstein, A., Kramer, J., Nuseibeh, B. (Eds.) 1994. *Software Process Modelling and Technology*, Taunton, Somerset, England: John Willey and Son Inc., Research Study Press, vol. 1.

Frankel, R. 1990. Solaris open windows the tooltalk service. Technical report, SunSoft.

Georgakopoulos, D., Hornick, M., and Sheth, A. 1995. An overview of WORKFLOW management: From process modeling to workflow automation infrastructure. *Distributed and Parallel Databases*, 3:119–153.

Grudin, J. 1994. Computer-supported cooperative work: History and focus. *IEEE Computer*, pp. 19–26.

Gruhn, V. 1994. Business process modeling in the WORKFLOW-management environment Leu. *Proc. of the Entity-Relationship Conference*, Manchester, UK.

Harel, D. 1987. Statecharts: A visual formalism for complex reactive systems. *Science of Computer Programming*, 8:231–374.

Harel, D. 1988a. STATEMATE: A working environment for the development of complex reactive systems. *Proc. of the 10th Int. Conf. on Software Engineering*, pp. 396–406.

Harel, D. 1988b. On visual formalisms. *Comm. of the ACM*, 31(5):514–530.

Harel, D. and Gery, E. 1996. Executable object modeling with statecharts. *Proc. of the 18th Int Conf. on Software Engineering*, Berlin.

Jacobson, I. 1992. *Object-Oriented Software Engineering: A Use Case Driven Approach*. ACM Press/Addison-Wesley.

Jones, T.C. 1986. *Programming Productivity*. New York: McGraw-Hill.

Junkerman, G., Peuchel, B., Schaefer, W., and Wolf, S. 1994. MERLIN: Supporting cooperation in software development through a knowledge-based environment. In A. Fikelstein and J. Kramer (Eds.), *Software Process Modelling and Technology*, John Willey and Sons, vol. 1.

Kaiser, G.E., Feiller, P.H., and Popovich, S. 1988. Intelligent assistance for software development and maintenance. *IEEE Software*, 5(3).

Malone, T.W., Lai, K., and Fry, C. 1992. Experiments with oval: A radically tailorable tool for cooperative work. *Proc of the ACM Conference on CSCW*, Toronto, Canada.

Marca, D.A. and Macgowan, C.L. 1988. *SADT: Structured Analysis and Design Technique*. New York: McGraw-Hill.

Mohan, C., Alonso, G., Gunthor, R., and Kamath, M. 1995. Exotica: A research perspective on workflow management systems. *Data Engineering*, 18(1).

Nuseibeh, B. and Finkelstein, A. 1992. Viewpoints: A vehicle for method and tool integration. *Proc. of 5th Int. Workshop on CASE*, Montreal, Canada: IEEE CS Press, pp. 50–60.

Otte, R., Patrick, P., and Roy, M. 1996. *Understanding CORBA the Common Object Request Broker Architecture*. Prentice Hall.

Paoli, F.D. and Tisato, F. 1994. CSDL: A language for cooperative systems design. *IEEE Transactions on Software Engineering*, vol. 8–20.

Promoter Group, (Ed.) 1996. *Software Process Technology*.

Purtilo, J.M. 1994. The polylith software bus. *ACM Transactions on Programming Languages and Systems*, 16(1):151–174.

Reinhard, W., Schweitzer, J., and Volksen, G. 1994. CSCW tools: Concepts and architectures. *IEEE Computer*, pp. 28–36.

Reiss, S.P. 1990. Connecting tools using message passing in the field environment. *IEEE Software*, 7(4):57–66.

Rombach, H.D. 1991. Practical benefits of goal-oriented measurement. In N. Fenton and B. Littlewood (Eds.), *Software Reliability and Metrics*, London: Elsevier Applied Science.

Roseman, M. and Greenberg, S. 1992. GroupKit: A groupware toolkit for building real-time conferencing applications. *Proc. of the ACM Conference on CSCW*, Toronto, Canada.

Rumbaugh, J., Blaha, M., Premerlani, W., Eddy, F., and Lorensen, W. 1991. *Object-Oriented Modeling and Design*. Prentice Hall.

Selic, B., Gullekson, G., and Ward, P.T. 1994. *Real-Time Object-Oriented Modeling*. Wiley Professional Computing.

Sutton, S.M. 1993. Opportunities, limitations, and trade-offs in process programming. *Proc. of the 2nd Int. Conf. on the Software Process*, Berlin, Germany.

Sutton, S.M., Heimbigner, D., and Osterweil, L.J. 1990. Language constructs for managing change in process-centered environments. *Proc. of the 4th ACM Symposium on Software Development Environments*, Irvine, CA. In it *ACM Software Engineering Notes*, 15(6):206–217.

Sutton, S.M., Ziv, H., Heimbigner, D., Yessayan, H., Maybee, M., Osterweil, L.J., and Song, X. 1991. Programming a software requirements-specification process. In M. Dowson (Ed.), *Proc. of the First Int. Conf. on the Software Process*, Redondo Beach, CA: IEEE Computer Society Press, pp. 68–89.

Verlage, M. 1994. Multi-view modeling of software processes. In B. Warboy (Ed.), *Proc. of European Wokshop on Software Process Technology EWSPT3*, volume 635 of LNCS, Villard de Lans, France: Springer-Verlag, pp. 123–127.

Yu, E. and Mylopoulos, J. 1993. An actor dependency model of organizational work—With application to business process reengineering. *Proc. Conference on Organizational Computing Systems (COOCS)*, Milpatas, CA.

Yu, E. and Mylopoulos, J. 1994. Understanding why in software process modelling, analysis, and design. *Proc. of the 16th IEEE Int. Conf. on Software Engineering*, Sorrento, Italy.

Automated Software Engineering 5, 97–132 (1998)

Federating Process-Centered Environments: The Oz Experience*

ISRAEL Z. BEN-SHAUL issy@ee.technion.ac.il
Technion-Israel Institute of Technology, Department of Electrical Engineering, Technion City, Haifa 32000, Israel

GAIL E. KAISER kaiser@cs.columbia.edu
Columbia University, Department of Computer Science, New York, NY 10027, USA

Abstract. We describe two models for federating process-centered environments (PCEs): homogeneous federation among distinct instances of the same environment framework enacting the same or different process models, and heterogeneous federation among diverse process enactment systems. We identify the requirements and consider possible architectures for each model, although we concentrate primarily on the homogeneous case. The bulk of the paper presents our choice of architecture, and corresponding infrastructure, for homogeneous federation among MARVEL environment instances as realized in the OZ system. We briefly consider how a single MARVEL environment, or an OZ federation of MARVEL environments, might be integrated into a heterogeneous federation based on ProcessWall's facilities for interoperating PCEs.

Keywords: collaborative work, distributed system, enterprise-wide environment, geographical distribution, internet, process interoperability, software process, workflow management

1. Introduction

Large-scale software engineering projects are not always confined to a single organization (e.g., group, department or lab), or even to a single institution (e.g., in a subcontracting or consortium relationship). A project may span *multiple* teams located at geographically dispersed sites connected by a wide area network (WAN) such as an organizational intranet or the Internet. Distinct teams may each have their own software development practices, favored tools, use different programming languages, etc. Yet the teams may still need to collaborate frequently in real-time, i.e., operate concurrently rather than sequentially, share part or all of their code and document base, perform tasks on behalf of each other and/or jointly, and so on.

Note we generally use the term "site" to mean an administratively cohesive domain, in which most (but not necessarily all) machines share a single network file system name space,

*This paper is based on work sponsored in part by the Defense Advanced Research Project Agency under [D]ARPA Order B128 monitored by Air Force Rome Lab F30602-94-C-0197, in part by National Science Foundation CCR-9301092, and in part by the New York State Science and Technology Foundation Center for Advanced Technology in High Performance Computing and Communications in Healthcare NYSSTF-CAT-95013. Part of this work was conducted while Dr. Ben-Shaul was a student at Columbia University. The views and conclusions contained in this document are those of the authors and should not be interpreted as representing the official policies, either expressed or implied, of the US or NYS government, DARPA, Air Force, NSF or NYSSTF.

e.g., cs.columbia.edu, as opposed to either a single host such as westend.psl.cs.columbia.edu, a lab subnet within an administrative domain such as psl.cs.columbia.edu, or a campus backbone such as columbia.edu. However, as we shall see, we sometimes use the term "site" in an alternative sense where a single local area network (LAN) or even a single machine may be home to multiple sites—when multiple teams happen to do their work on that same LAN or machine, respectively. That is, a site is wherever a team does its work.

Consider, for example, several teams each responsible for a separate set of "features", all intended to be included in an upcoming Microsoft product release (Cusumano and Selby, 1995). Imagine some of these teams have been subcontracted from various independent software houses located outside Microsoft's main development lab, perhaps even outside the United States. Although Microsoft documents recommend vendor processes, it seems unlikely that these teams would follow identical software engineering practices, use exactly the same tools, etc. They may not be willing to publicize (even among themselves) their proprietary software development "trade secrets".

There are various approaches to software development environment (SDE) support for multi-site projects. For the purposes of this paper, we organize these approaches along two orthogonal axes: tightness of coupling and degree of heterogeneity. At one end of the coupling spectrum, each team chooses its own SDE (which may happen to be a copy of the same environment in some of the sites) and there may be more or less concern with whether the different team's SDEs are compatible with each other.

A little further along the coupling spectrum, the teams may choose the same (homogeneous) SDE, to minimize data conversion and supply a common vocabulary, or they may use heterogeneous SDEs but agree on a shared data interchange format. In either of these cases, sharing and collaboration between teams is done *outside* the environment—unless some special "glue" is added on top to bind them together into a federation (i.e., a common data format alone is not sufficient for them to work together at run-time), as explained below.

Another important intermediate range is covered when the teams share the same instance of what we call a *multi-site* SDE, which distinguishes among teams (who may reside at the same or different sites) in some way, but provides facilities for sharing and collaboration between teams *inside* the environment. That is, the glue (or perhaps "cement" in this case) is part of the environment framework itself. The degree of independence afforded by each team determines the point within the subrange. The heterogeneous version of this intermediate range consists of *multi-SDEs*, that is, interacting but distinct SDEs with the glue consisting of a shared standard event notification scheme (Barret et al., 1996) or other control facilities in addition to a common data interchange format.

Finally, the far extreme is a geographically distributed SDE that does not distinguish among teams—all the users are treated as members of one very large team sharing everything. We choose the terms "multi-site" and "geographically distributed" here because many SDEs are said to be "distributed", meaning they have multiple internal components that may execute on different hosts on a LAN or WAN.[1]

The geographically distributed SDE end of the spectrum is analogous to distributed database systems, where there is transparent access to distributed data, while the independent choice of SDE end is comparable to a collection of independent databases. The database community has also delineated an intermediate range, often termed "federated databases" (Sheth and Larson, 1990; Ram, 1991). Federated databases generally permit a high degree of autonomy with respect to one or both of two criteria, schema and system: local components of a single database system with intrinsic federation glue may devise and administer their own schema independently (known as a *homogeneous federation*), and/or the local components may correspond to different database systems from among those supported by extrinsic federation glue (*heterogeneous federation*)—in which case even conceptually equivalent schemas may appear in different forms due to system-specific data definition languages.

We are concerned in this paper with the subclass of SDEs known as *process-centered environments* (PCEs) (Schäfer, 1996; Boehm, 1996). In general, a PCE is a generic environment, framework, or kernel, intended to be parameterized by a *process model* that defines the software development process for a specific instance of the environment. The PCE's *process engine* interprets, executes, or "enacts" the defined process, to assist the users in carrying out the process by guiding them from one step to another, enforcing the constraints and implications of process steps as well as any sequencing or synchronization requirements, and/or automating portions of the process. A federated PCE might coordinate users from multiple teams working on collaborative tasks, inform one team when it should perform some task on behalf of another, notify one team on completion of some task it has been waiting for another to perform, and transfer process state and product artifacts (design documents, source code, executables, test cases, etc.) among local components of the federation as needed for this work.

It is important to note that in both multi-site PCEs and multi-PCEs, we treat process as the *integrating principle* of federation. That is, the federation is intended to fulfill the semantics expressed explicitly in the (global) process, and this has a crucial impact on the design of the federated architecture. We do not address non-process-centered SDE federations further in this paper.

A multi-site PCE is analogous to a homogeneous database federation. In particular, the PCE process model fills the role of the database schema with respect to homogeneous federation: the local components of the multi-site PCE are identical, except that they are tailored by and thus enact different process models (or possibly reflect different instantiations of the same process model). A multi-PCE is analogous to a heterogeneous database federation, and similarly requires that each separate PCE is independently (from the others) capable of interfacing to federation glue that makes it possible for them to work together. Generally the process modeling formalism as well as the process model are different at each site, although the process model could be conceptually the same while expressed differently. In either case, again, the process model fills the role of the database schema, although we note that generally each PCE also supports some schema for a data repository containing its software development artifacts and process state.

Figure 1 illustrates the space of approaches, highlighting the two "federated" grey areas, which serve as the context of this paper. That is, we are concerned with federated PCEs that exhibit at least some degree of coupling but also at least some degree of independence, i.e., not transparent distribution; and we do not consider completely homogeneous approaches, where even the processes must be identical, or completely heterogeneous approaches, where

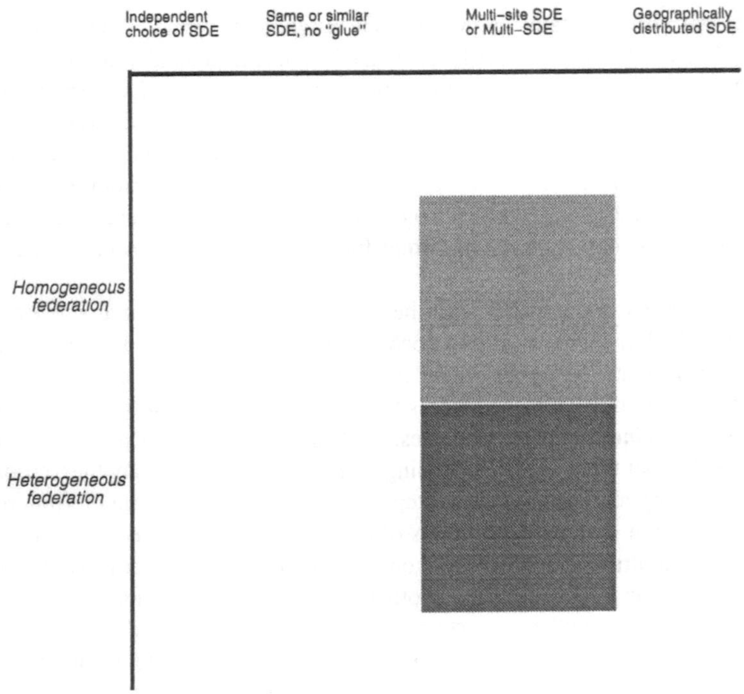

Figure 1. Multiple team PCE spectrum.

it is impossible to introduce any sort of run-time integration, and the only possible integration is at definition-time, through a process definition exchange format, e.g., as promoted by WfMC (Workflow Management Coalition Members, 1995).

A federated PCE for cross-organization projects should permit each team to specify its own local process model, along with the desired collaboration with other teams through shared subprocesses, tool sets, data subschemas, data instances, etc. Thus, a noticeable difference from database federation is that the focus here is on interoperability among heterogeneous processes, i.e., the application semantics, as opposed to (only) heterogeneous data schemas, i.e., the data on which applications operate.

One approach to homogeneous PCE federation, where every team runs a component of the same multi-site PCE but enacts a different process, is taken by our OZ PCE (Ben-Shaul and Kaiser, 1994b). OZ was devised to scale up our earlier MARVEL PCE (Kaiser and Feiler, 1988) to multi-process, multi-team, geographically dispersed software engineering projects. OZ introduces an *International Alliance* metaphor whereby each team autonomously devises its own local process (supported by a local OZ component that is essentially an extended instance of MARVEL), analogous to how each country has its own local customs and laws. A team may agree to extend its process to a small degree (and thus temporarily lose some autonomy) in order to participate in a *Treaty* with one or more other teams. The enactment of a *multi-site task*, defined as any task that involves interaction among the several sites of a multi-site PCE, is called *Summit*. OZ extends MARVEL with Treaties, Summits, and an underlying inter-site communication and configuration infrastructure where each site corresponds roughly to an instantiated MARVEL environment.

Only tasks specified in a Treaty may access data from other sites, and, even then, only in accordance with the privileges granted by the Treaty. For example, a site may agree to perform certain tasks requested by another site on its own local data; or a site may agree to allow another site to perform certain tasks on its local data; or a site may agree to perform certain tasks on data from several sites. However, each site (or team) is responsible for any prerequisites or consequences of such tasks with respect to its own data, following its own process, just as in preparation for and follow-up of meetings among country leaders (the basis for our metaphor). Treaties may be dynamically defined while the process unfolds, i.e., while computation is in progress, permitting a degree of flexibility not found in most distributed systems.

One approach to heterogeneous PCE federation, where two or more distinct process systems are bound together into a multi-PCE, is taken by Heimbigner's ProcessWall (Heimbigner, 1992). Note that Process Wall is the external glue supporting such binding, *not* itself a PCE. ProcessWall could of course be used to integrate multiple instances of the same PCE, say MARVEL, with different process models as in OZ, but in this paper we address only the more challenging case of using it to federate multiple distinct PCE systems.

Heimbigner refers to ProcessWall as a *process state server* because it enables interaction between the PCEs through a centralized representation of global process state that the teams agree to share. However, we believe it is more useful to treat the mechanism Heimbigner describes as a *process task server*: it may maintain the history of tasks that have already been completed, in aggregate representing the current process state, but more significantly from the viewpoint of federation the server posts those tasks that have been instantiated but not yet scheduled for enactment by one of the participating PCEs.

In particular, each participating PCE manages, schedules, and enacts its own task descriptions, usually forwarding each description to ProcessWall only after that task has been completed, e.g., to allow users to exploit ProcessWall's process state inspection facilities (part of the glue). Thus, the process remains primarily decentralized, since the actual process operation is performed by the separate PCEs without any interactions between them or with ProcessWall while the work is in progress. However, in some cases a PCE may send an instantiated (e.g., with data parameters) but unenacted task to ProcessWall intending it to be executed by some other PCE in the federation, because the sending PCE does not have the data, tool(s) or user(s) appropriate to conduct the work.

An intelligent scheduler might then be attached to ProcessWall to direct such posted tasks to particular sites, as described in (Osterweil, 1995), or alternatively ProcessWall might be treated as a "blackboard" (using artificial intelligence terminology (Hayes-Roth, 1985)) from which the schedulers of the individual PCEs participating in the federation select those tasks they are suited to perform. Any sharing of software product artifacts, as opposed to process state, is implicit in the data information included with posted tasks. As in OZ, each site might autonomously devise its own process model.

Mentor (Weissenfels et al., 1996) is similar to ProcessWall but divides the process state/task server into two components: a *worklist manager* acting as a pure task server and a *history manager* corresponding to a pure state server; data sharing is factored out as in ProcessWall. Note Mentor is a workflow management system intended for business applications, not a PCE oriented towards software engineering; whether there is any fundamental difference between workflow and process is a matter of some debate (Sheth et al., 1996), but we blur the distinction in this paper. In any case, heterogeneous federation based on Mentor would probably be quite similar to the ProcessWall model.

Process interchange formats (Workflow Management Coalition Members, 1995; Lee and Yost, 1994) support translation of a logically single process model into the different representations of distinct process systems, but do not provide any means for collaboration and interoperability during the process enactment by those systems. Thus, there is no true federation in the sense addressed by this paper. However, some kind of translation facilities are needed as part of any heterogeneous federation: Mentor transforms the heterogeneous process modeling formalisms into StateMate charts (Humphrey and Kellner, 1989), but in the case of ProcessWall only process state is translated (or the participating PCEs might be implemented to use a common task format).

We mentioned above that process enactment by a federated PCE might involve movement of product artifacts among teams that could potentially be distributed across a WAN. Alternatively, all the sites might share a common centralized data repository, presumably located at one of the sites, or even a transparently distributed data repository. Globally shared data seems most appropriate for projects organized far in advance and involving only a single institution, perhaps with multiple campuses. In contrast, when different institutions work together, particularly when the federations are dynamically created and dissolved, most likely the institutions would prefer to maintain locally at least those product artifacts produced by their local process.

This paper discusses the architectural aspects of PCE federation and associated infrastructures, and then justifies our architectural choices for the fully implemented (and used in our day-to-day work since April 1995) OZ system in detail. We also explore a hypothetical OZ/ProcessWall interface. Our investigation of architecture is strongly influenced by the fact that the main purpose of PCE federation is to enact multi-site and global processes. For example, global processes devised using a top-down methodology, say, intended for multiple campuses of a single institution, may require somewhat different architectural support than global processes constructed in a bottom-up manner, e.g., for temporary multi-institution collaborations. However, methodologies for developing global processes are outside the scope of this paper.

First we present architectural requirements and the alternative architectural models we considered for both homogeneous and heterogeneous PCE federations, the latter in contrast to the former (i.e., many of the requirements are shared and the architectures are analogous). We then elaborate the specific design decisions and tradeoffs that were made in developing the OZ architecture and infrastructure that extended our earlier MARVEL PCE to a homogeneous PCE federation. We do not go into detail regarding ProcessWall, Mentor, or any other such heterogeneous federation glue, since that is properly left to their developers. Instead we briefly discuss how MARVEL, or OZ, might be integrated into a heterogeneous federation based on ProcessWall's process state/task server model, to some extent synergizing the two federation mechanisms, i.e., allowing OZ to operate as a multi-site PCEs within a multi-PCE. In both sections, the range of architectures is explored specifically in the context of our choices for Oz. We conclude with the contributions of this work and outline some directions for future research.

2. Requirements and alternative architectures

In both the homogeneous and heterogeneous federated models, each local site runs a component of a multi-site PCE or multi-PCE. We refer to such a site component as a *subenvironment*, or just SubEnv, even though it may operate in stand-alone fashion as a full PCE. We refer to the "glue" that holds the SubEnvs together as the federation's *foundation*, or just Foundation. Database federation involves a similar foundational component or layer, e.g., to control global transactions, although many classes of distributed system do not include any foundational layer beyond a basic networking communication protocol. This section of the paper is concerned with the functionality (Section 2.2 for the homogeneous case and Section 2.5 for the heterogeneous case) and architectural design (Sections 2.3 and 2.6, respectively) of the Foundation. Recall that we are mainly concerned with multiple sites on a WAN, generally with independent administrative domains—although of course nothing prevents multiple sites from running on the same LAN, that is, a multi-site PCE or multi-PCE might operate entirely within a single organization or group and each "team" could conceivably consist of only one user (as in the Oz EmeraldCity environment we use to support our own software development (Kaiser et al., 1997)). We take as given the requirement that each site must be able to support an autonomously devised process model.

2.1. Local environment internal architecture

Although the focus of this paper is on *federation* architectures, it is useful to begin the discussion with an overview of SubEnv *internal* architectures, since they have a substantial impact on the design of a homogeneous federation; internal architecture is less germane in the case of heterogeneous federation since, in general, each participating SubEnv may employ a different internal architecture. As we focus in this paper on process-centered SDEs and the impact of process on architecture, we characterize local PCE architectures based on the degree of centralization in process enactment, comprised of two aspects: process control and tool execution. The former refers to the function of deciding which task to enact and when, according to process constraints/context, whereas the latter corresponds to where and how the task gets executed, often but not necessarily via one or more specific tools. This separation is important in PCEs because it reflects the typical separation between the process itself and the tasks spawned by it, which may invoke external tools, take arbitrarily long to complete, involve one or more human (possible simultaneous) users, and so on.

Although our goal was to scale up our pre-existing MARVEL PCE to support multiple teams each sharing a potentially different process, where the teams might be connected by either a LAN or a WAN, we identified four classes of internal PCE architecture—only one of which applies to the final MARVEL version 3.1.1 we were concerned with. Note these are *not* the same classes suggested by Peuschel and Wolf (1993) and we follow a different classification scheme: Peuschel and Wolf were concerned with the relationship between the process engine and the data repository, whereas we consider process control vs. tool (or task) execution.

1. **Centralized process control and centralized tool execution.** This is the simplest case, where both control and execution are carried out by the same component. An all-in-one single-user PCE such as MARVEL 2.x (Kaiser et al., 1990) and some compiled process programs, e.g., written in APPL/A (Sutton et al., 1995), would fit into this class. Even a client/server system might fall into this category if the client supported only the user interface and all process enactment was performed in the server. Given the multi-user multi-task nature of practical software engineering processes, this architecture is inherently unscalable, even for a single team.
2. **Centralized process control and decentralized tool execution.** A process server maintains the state of the process, controls its enactment, and synchronizes access to shared resources, but the tools themselves execute at process clients. MARVEL 3.x (Ben-Shaul et al., 1993), Process WEAVER (Fernström, 1993), and Mentor fit this mold, albeit in different ways. MARVEL 3.x relies on fixed user clients to fork tools, whereas Process WEAVER spawns user "work contexts" as needed by the process. OZ local sites are somewhere in between, with one server per site (i.e., per team), generally employing user clients as in MARVEL 3.x but also supporting "proxy clients" that run tools on behalf of one or more users under various circumstances, as explained in (Valetto and Kaiser, 1996; Skopp, 1995). Mentor is similar to OZ in that user clients can connect to multiple process servers in the federation.
3. **Decentralized process control and centralized tool execution.** Control is distributed among multiple process servers, where the tool execution function is supported by a single component. This model supports separate process engines for each user—as in Endeavors (Bolcer and Taylor, 1996) or Merlin (Schäfer et al., 1992)—while sharing special computational or database resources used in tool invocation. One can easily imagine multiple workflows accessing the same tool management resource, particularly if only the tool broker is centralized, directing actual tool invocation to distributed hosts as in WebMake (Baentsch, 1995).
4. **Decentralized process control and decentralized tool execution.** Here the process itself is distributed across multiple nodes, where each node is responsible for the execution of its subprocess as well as corresponding tasks. Control flow and synchronization between the process segments is specified locally inside the nodes. Several transactional workflow systems, such as Exotica (Alonso et al., 1995) and Meteor (Jin et al., 1993), operate in a fully distributed manner—by expressing the workflow implicitly in a network of task managers (which invoke the actual tools) that interact only with their predecessors and successors in the workflow routing.

Several issues influence the choice of single-site PCE architecture. A major factor is the level of data integration employed by the PCE for product artifacts. PCEs with extensive data integration facilities (e.g., SPADE (Bandinelli and Fuggetta, 1993), EPOS (Conradi et al., 1991)) might choose a centralized control architecture to minimize communication between the data and process managers when disseminating tasks—unless the data management is itself distributed, and/or the data itself is physically distributed, in which case a distributed control architecture may be employed. PCEs with no data integration facilities might be fully distributed in an easier manner. Note, however, that full distribution of process enactment is not incompatible with sharing a centralized data repository; see (Peuschel and Wolf, 1993).

Another characteristic that impacts the choice of local PCE architecture is whether the process modeling paradigm employed by the PCE is reactive or proactive (termed proscription vs. prescription by Heimbigner (1990)). Reactive enactment may be realized better in a centralized-control architecture, as requests for enactment are directed to a single server that dispatches the service to a client (perhaps the requester itself), whereas proactive enactment may be distributed by assigning a priori each task to a component, with the ordering and execution constraints inside each component—or implicit in their interconnection topology.

2.2. *Requirements for homogeneous federation*

We have identified the following additional requirements for the homogeneous model:

- **Communication infrastructure.** The most fundamental functional requirement for multi-site process enactment is that the Foundation include an infrastructure whereby the SubEnvs communicate with each other regarding multi-site tasks. This might be constructed directly on top of TCP/IP sockets, or employ some higher level mechanism such as RPC or CORBA (The Common Object Request Broker: Architecture specification revision 2.0, 1996). In any case, we are mostly concerned with the PCE-cognizant interconnectivity layer, i.e., the Foundation, not the underlying mechanism.
- **Global process definition and acceptance.** On top of the basic interconnectivity support, the Foundation must supply means for local processes to interoperate, i.e., to model and enact tasks that in some way span multiple processes/sites and contribute to the global process. In particular, the Foundation must have facilities to (re)negotiate and (re)define (possibly dynamically) the specifics of process-interoperability for the relevant processes, e.g., via Oz-like Treaties.

  Although a degenerate global process may involve only primitive operations (e.g., copy a data item), we, in general, assume some notation to define multi-site tasks whose enactment is controlled to some degree by the Foundation. In other words, we assume that multi-site tasks are themselves modeled in either top-down or bottom-up fashion as parts of a global process, with conceptually its own state and purpose. However, multi-site process modeling and enactment is the subject of another paper (Ben-Shaul and Kaiser, 1995a); here we are concerned with structure and organization of components, i.e., architecture, that supports multi-site processes.
- **Local autonomy and independent operation.** As far as purely local work is concerned, i.e., work involving the local process operating only on local data, a SubEnv should operate autonomously and independently, and provide the same capabilities as would a single-site PCE. It should not in any way rely on communication with other SubEnvs, or with the Foundation, in order to perform its standard functions with regards to defining and executing the local process. The underlying assumption is that most of the work done

by a site is local to that site, and therefore the multi-site PCE should still be optimized towards local work.

- **Restrict global dependencies to affected sites.** A related issue is that the SubEnv should minimize the dependencies on uninvolved SubEnvs when executing part of a multi-site task. These two requirements are somewhat similar to control and execution autonomy, respectively, in multi-database transaction management (Korth, 1986). The local site autonomy prized in the Oz approach to bottom-up process modeling has also been argued as necessary for top-down modeling: "A participant on a lower level (of the hierarchy) does not want his/her management to know how a task is performed" (Schwenkreis, 1996). Thus we rationalize site autonomy as a critical requirement.
- **Configuration awareness.** The SubEnvs must somehow be aware a priori (statically), or become aware during the course of process enactment (dynamically), of each other's existence, i.e., the other members of the currently configured federation, if they are intended to directly communicate and, possibly, collaborate. In the case where a Foundation intermediary is the conduit for all communication and interactions among SubEnvs, the SubEnvs must at least be aware of that intermediary and vice versa.
- **Dynamic (re)configuration.** Since the lifetime of enacted processes is often long, months to years, the Foundation must allow for SubEnvs to join or leave a federation while a process is in-progress, that is, support configuration and reconfiguration of participants in the global process. It is of course also necessary for SubEnvs to determine or negotiate what services each can expect from other (perhaps anonymous) SubEnvs in terms of process control, tool execution, and data and other resources, and how to coordinate exploitation of those services, but again that is the subject of another paper (Ben-Shaul and Kaiser, 1995a).
- **Process state survives failures.** Since processes in general, and federated processes in particular, are enacted for long durations, they require facilities for persistent process state. In cases where each local PCE manages its own product-data repository, the Foundation must also provide mechanisms for transferring product artifacts, in addition to process state, among sites. This may involve the same or different inter-PCE communication channels for product vs. process data, but the two cases have to be handled separately because products typically involve significantly larger volumes of data. For example, in a multi-site `build` task, one site may collect code modules from the other relevant sites and return to them copies of the resulting executables and/or libraries. Another example is a distributed groupware task such as multi-user editing, in which source code and/or documentation files stored at one site may need to be (simultaneously) transferred to several other sites. In general, bulk data may be temporarily cached, permanently copied, or migrated between sites.
- **Semantics-based transaction management.** Another data-related requirement involves support for sophisticated and flexible concurrency control and failure recovery mechanisms due to the long duration of tasks and task segments, interactive control by users, and human-oriented collaboration among tasks and task segments while they are in progress (Korth, 1995). The explicitness of the process in PCEs makes it possible to employ semantics-based transaction management (Barghouti, 1992; Heineman and Kaiser, 1997). Multi-site tasks may modify data from multiple sites, and thus require

some kind of global transactional support, such as two-phase commit that interfaces with local transaction managers. Investigation of this topic is beyond the scope of this paper, see (Ben-Shaul and Heineman, 1996; Heineman, 1996).

2.3. Homogeneous federation architectures

We identify five categories of architectures within the homogeneous (light grey) area of the "tightness of coupling" spectrum of figure 1. Note relatively minimal (or no) translation services are needed in any of these categories: all the SubEnvs speak the same languages (including data formats, process modeling notation, and tool wrapper scripts). The Foundation may perform name mappings, since a common ontology is not assumed, but this is not its major function.

For the sake of the figures depicting the multi-site PCE architectures below we show one plausible set of components that may comprise a local SubEnv (process, data, tool, and user interface components), but we do not intend in any way to specify or constrain a SubEnv to follow the given structure, and any of the internal architectures discussed in Section 2.1 might be employed. Further, we do not specify any particular internal component for interfacing to the Foundation—interpret the figures as if they *all* (potentially) do, to achieve federation of their various functionalities.

2.3.1. Ad hoc. Two or more instances of the single-site PCE are hardwired together in some ad hoc fashion for a particular purpose. There is generally no Foundation, per se. This model obviously does not scale, so is not addressed further.

2.3.2. Centralized glue. The SubEnvs communicate and interact through a single centralized glue component that constitutes the Foundation. As mentioned earlier, the Foundation, i.e., the federation glue, is intrinsically part of the multi-site PCE rather than imposed externally. However, each SubEnv necessarily includes code to interface to the Foundation, perhaps through RPC or TCP/IP socket calls originally part of native SubEnv if the PCE was designed as a multi-site PCE, or inserted later if not. The Foundation may perform brokerage or routing among SubEnvs, and maintain the state of multi-site process segments.

Figure 2 illustrates this architecture. Distributed systems with one or more central components do not scale beyond a certain level, since the centralized component becomes a performance bottleneck and single point of failure (i.e., if this one component fails multi-site tasks become impossible). The interface aspect of the centralized glue could be expanded in several different ways with respect to the SubEnvs, analogous to the intermediary, moderated and direct decentralized cases below. We do not discuss these options, since the variations between the cases are overwhelmed by Foundation centralization—although as the interfaces get "larger" the central component tends to get "smaller", as functionality is shifted, effectively achieving a hybrid between centralized and decentralized approaches.

2.3.3. Decentralized glue. The SubEnvs communicate and interact through *intermediaries*, with one intermediary attached to each SubEnv. These intermediaries collectively

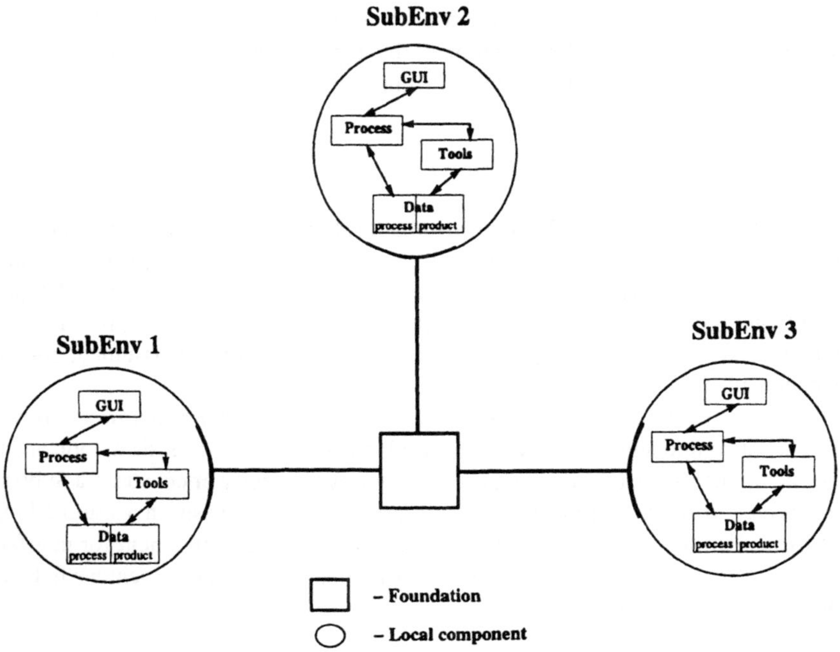

Figure 2. Centralized architecture.

constitute the Foundation glue, and there is *no* centralized component. A Foundation intermediary may or may not be realized as a separate operating system process from its local SubEnv. If separate, it would usually reside "close" to the local SubEnv, e.g., on the same LAN, but not necessarily on the same host.

However, this case is distinguished from the peer-to-peer cases below in which the intermediary has no special knowledge of the PCE's process-oriented functions and no access to its process model, nor any special knowledge of its tool execution facilities, data repository, user interface, etc. To the degree that these internal functionalities (whether or not distinguished as components) interoperate within the federation, and thus interact with the Foundation infrastructure, they must interface to the intermediary. The intermediaries are tightly coupled with each other, e.g., maintaining long-term connections which permit them to share the Foundation's global process state and work closely together to realize the Foundation's functionality (e.g., a distributed name service), but are loosely coupled with respect to their SubEnvs. See figure 3. From a process perspective, the interaction between the global process and the local processes is quite limited because the Foundation has no access to the internals of the local processes.

Note a geographically distributed realization of the decentralized glue architecture is plausible—with the intermediaries acting as gateways to remote SubEnvs on a WAN; the two peer-to-peer architectures below also easily admit a geographically distributed implementation. Although a centralized architecture might also be geographically dispersed, this seems less likely from an administrative point of view—except possibly within a organizational

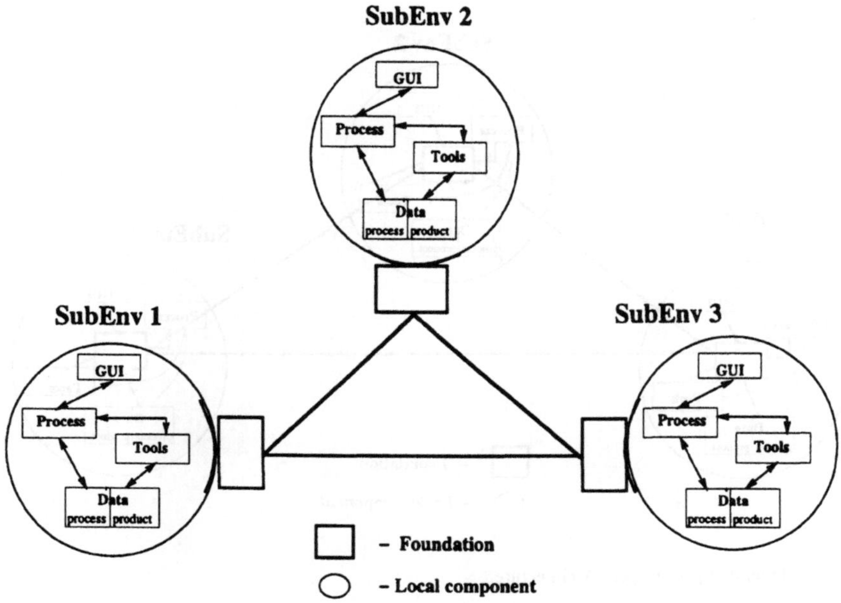

Figure 3. Decentralized glue architecture.

intranet where the same organization owns and controls all the relevant sites including the machine hosting the central component.

2.3.4. Moderated peer-to-peer. The SubEnvs again communicate/interact through intermediaries, which we call *moderators* here, with one moderator attached to each SubEnv. These moderators collectively constitute the Foundation, and, again, there is *no* centralized component. Again, there is no implication intended regarding physical realization, the moderator may or may not be realized as a separate operating system process from the rest of its local SubEnv. If separate, again it would necessarily reside "close" to the local SubEnv, most likely on the same host.

Unlike the decentralized glue case, here each moderator is tightly coupled with its local SubEnv and has intimate knowledge of that SubEnv's process-oriented expectations regarding services from other SubEnvs. Similarly, the moderator is cognizant of the local process model and state, tool execution, data repository, user interface, etc., if relevant to federation. Again, to the degree that these internal functionalities (whether or not distinguished as components) interoperate within the federation, and thus rely on the Foundation infrastructure, they must interface to the moderator. In contrast, the moderator is loosely coupled with respect to its peer moderators, e.g., making only short-term stateless connections. See figure 4.

This approach, again, seems obviously more likely to scale than a centralized architecture, but more-or-less equivalent with respect to scaling as the decentralized glue case. However, in this case the architecture cannot assume any shared capabilities (e.g., name services) provided by the Foundation. In other words, it is a "shared nothing" architecture as far as the

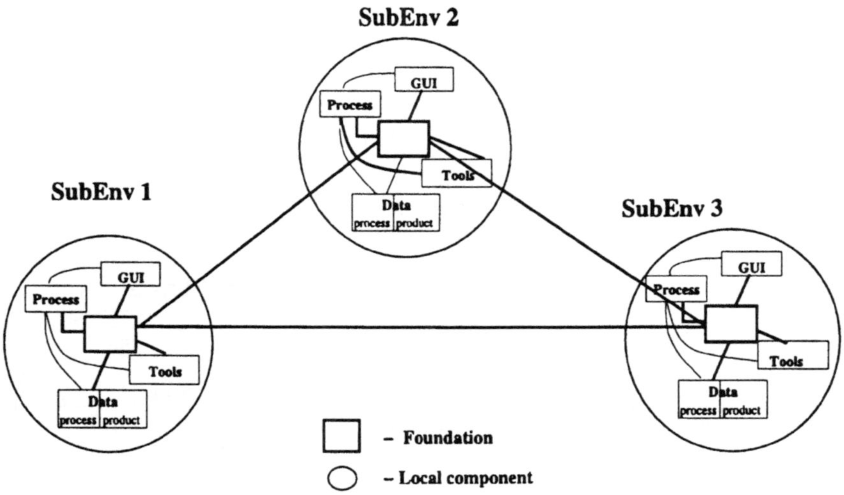

Figure 4. Moderated-peer-to-peer Architecture.

Foundation is concerned. (Note this does not preclude sharing among internal components of the SubEnv.) On the other hand, the interaction between the global process and the local processes is richer, because the Foundation has direct access to local processes.

That is, the primary distinction between the decentralized glue case and the moderated peer-to-peer case is that in the former the local Foundation components have *no* knowledge of the local processes and manage a multi-site process imposed on the local SubEnvs divorced from their local processes, whereas in the latter the local Foundation components have intimate knowledge of the local processes, but without any shared global process state or common control. This reflects the tradeoff between stronger coupling within the Foundation and weaker coupling between the local component of the Foundation and its local SubEnv, in the decentralized glue case, vs. weaker coupling within the Foundation and stronger coupling between the local Foundation and its SubEnv, in the moderated peer-to-peer case.

2.3.5. Direct peer-to-peer. The SubEnvs communicate/interact with each other directly, and the Foundation cannot easily be distinguished from the rest of the multi-site PCE. That is, the local component of the Foundation is *built into* one or more of the SubEnv's internal components, most likely the process engine; there is no specific component introduced *solely* to represent the Foundation infrastructure. See figure 5.

While this approach probably offers improved performance over the others described above, it is more challenging to realize for pre-existing single-site PCEs because it generally involves significant modification throughout the PCE code as opposed to "adding on" interfaces to a new component. Thus, scaling is restricted for software engineering rather than distributed computing reasons.

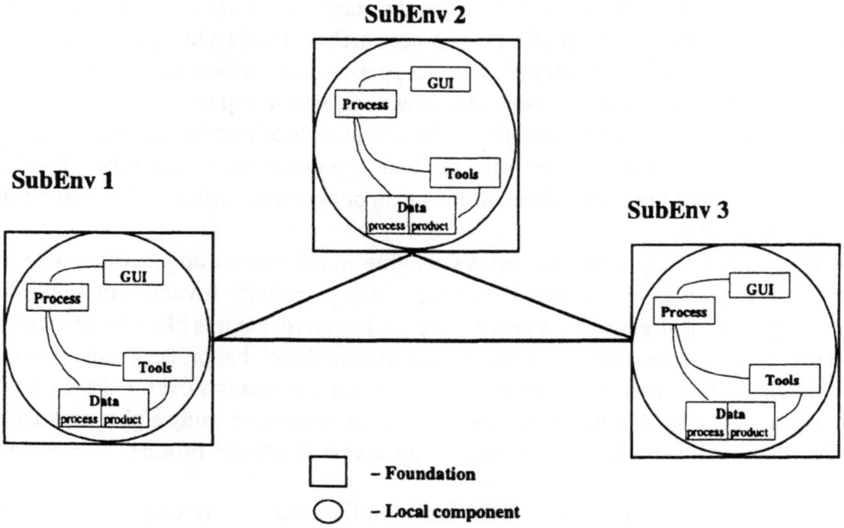

Figure 5. Direct peer-to-peer architecture.

2.4. *Choice of homogeneous architecture*

The choice of federation architecture for homogeneous multi-site PCEs depends largely on two concerns:

1. The paradigm chosen for modeling and enacting federated processes.
2. The style, design, and implementation of the local SubEnv framework.

Regarding multi-site or global processes, we distinguish between two major paradigms, top-down and bottom-up, although of course hybrids are possible. *Top-down* refers to a process broken down through multiple levels of granularity each corresponding to subsequently smaller organization units, as in the enterprise-level to campus-level to department-level to group-level of the Corporation metaphor (Shy et al., 1989); this is analogous to a global transaction in federated databases (Pu, 1988). *Bottom-up* refers to interoperability among possibly pre-existing local processes, as in OZ's International Alliance metaphor, without a global overseer (unfortunately, the kind of distributed computing scenario where the Byzantine Generals problem arises—although consideration of fault tolerance in the face of malicious behavior is outside the scope of this work). We do not consider here which of the two paradigms is more appropriate for various applications (see (Ben Shaul and Kaiser, 1995a) for such a discussion), but rather which *architecture* best supports each of the paradigms—particularly the bottom-up paradigm, since one of our major goals was to link pre-existing single-site MARVEL processes.

In order to support top-down global processes, the federation must support maintenance of global process state. This suggests a glue architecture, particularly centralized but also decentralized, where the Foundation manages the state. In contrast, bottom-up federation can naturally be realized on top of a peer-to-peer architectural style, again in one of two possible ways, namely, the moderated or direct peer-to-peer architectures. In other words, we make a primary distinction between top-down vs. bottom-up process interoperability, and a secondary distinction between the architectural realization of each style. In general, bottom-up interoperability is more scalable than top-down, as in any other distributed system, but introduces process-related problems in our context such as lack of explicitness of the global process.

The association of top-down processes with glue and of bottom-up processes with peer-to-peer architectures is not exclusive, however. It is potentially feasible, for example, to realize a top-down process using a peer-to-peer architecture, but it is likely to be inefficient and harder to realize because of the needs to distribute the global process state among the loosely coupled intermediaries and to manage shared information over a shared-nothing architecture. It is probably easier to realize a bottom-up process using a glue architecture, provided that administrative barriers (regarding access to private process state at remote sites) can be relaxed or overridden.

Let us now consider the impact of the local SubEnv architecture on the choice of federated architecture. One factor stems from the degree of openness and extensibility of the process control and tool execution components (the data repository is also of concern, but those issues are not terribly different than in other federated database applications, so we concentrate here on PCE-specific matters). In particular, peer-to-peer architectures demand tighter integration at the process control and task/tool execution levels, between the local SubEnv and its Foundation component, than glue architectures. This is possible only if the local SubEnvs provide suitable application programming interfaces (APIs) for extending these functionalities—which would usually suffice for moderated peer-to-peer. Or, alternatively, if the SubEnv source code can be internally modified—which would by definition be required for the direct-peer-to-peer, assuming the PCE was not originally implemented as a multi-site system. (We know of none that were, e.g., multi-site OZ was realized by adapting the single-site MARVEL 3.x PCE.)

Another important factor that impacts mainly peer-to-peer architectures is the degree of centralization of the SubEnv internal architecture. SubEnvs with centralized (local) process control naturally lend themselves to a direct peer-to-peer federation architecture, where the Foundation infrastructure is built into the local process engine—which becomes the conduit to communicate with other SubEnvs; communication via a centralized tool manager is also conceivable. Fully decentralized (local) process enactment, in contrast, seems better suited to a moderated peer-to-peer architecture since there is no one component that stands out as the focal point. Instead, a new moderator component is attached to the SubEnv as a whole and communicates with each of the other local components as well as with its peer moderators. However, a direct peer-to-peer architecture is not inconceivable for decentralized SubEnvs; see (Wallnau et al., 1996).

To summarize, the above categories represent different degrees of (de)centralization of the Foundation, ranging from a logically and physically centralized architecture, to several forms of logically and physically decentralized architectures with variations in the coupling between and within SubEnvs. Our key observation is that there is no one architectural style for federated PCEs that is inherently superior to all others. Instead, we argue that the choice of a proper architecture depends on the requirements of the system, and more specifically on the architecture of the local PCE and on the federated process paradigm. This is elaborated for the case of Oz scaling up MARVEL in Section 3.

2.5. *Requirements for heterogeneous federation*

Recall that in the heterogeneous model, each site (or team) runs a separate PCE that works together with other PCEs in a multi-PCE joined together via the Foundation. Each site may employ a *different* local PCE, selected to best fulfill its own needs or retained for historical reasons. A few may happen to use independent copies of the same system, or local PCEs with similar architectures and interfaces, but we cannot count on that and therefore treat each SubEnv as unique within its federation.

We have identified the following requirements for heterogeneous federation. In general these are *in addition to* homogeneous federation requirements, although in some cases we repeat the seemingly identical requirement followed by new discussion oriented towards the special circumstances of the heterogeneous case.

- **Communication infrastructure.** The most basic function of the Foundation is to communicate with each SubEnv participating in the federation. In general, the SubEnvs cannot communicate directly with each other since (by definition) they were designed as independent PCE systems (although perhaps following a "standard" Foundation interface). Realization of multi-site tasks, or fulfillment of a request from one site for another site to undertake a task on its behalf, requires that some logically homogeneous component is added to each PCE so that it can bind into the federation. This component may involve quite diverse per-PCE physical implementations, e.g., to perform PCE-specific protocol conversions and data format translations. The conceptually common component is relatively limited, though, and in particular does not take over local process modeling and enactment functions—since otherwise we could consider it to effectively convert the federation to the homogeneous case.
- **Federation awareness.** The SubEnvs (usually) must somehow be made aware of any federation(s) in which they participate, possibly more than one at a time, in order to interoperate and contribute to global process enactment. The SubEnvs need no direct knowledge of the other SubEnvs in the federation, per se, but there must be some means whereby the Foundation coordinates the global process, either by *notifying* a given SubEnv that it should or could perform specific tasks or by posting the request to some standard forum that each SubEnv *polls* to choose tasks it is able and willing to perform.

  Note this does not necessarily assume that SubEnvs have some means to inform the Foundation of pending tasks that they are unable or unwilling to do themselves: the Foundation could itself impose *all* tasks, perhaps through a special process modeling and enactment system intended to act as a "global hand" supporting some form of "superworkflow" (Schwenkreis, 1996), analogous to multi-part transactions submitted to heterogeneous multi-databases.

In principle, it might be plausible for a SubEnv to perform work on behalf of a federation without ever noticing that the heterogeneous federation exists (which would not normally be the case for homogeneous federations). Thus, an alternative is that only the Foundation is aware of the various SubEnvs, and picks up their results through some non-intrusive manner, such as understanding file formats of what the PCE considers internal process state information (as done, for example, in (Heineman and Kaiser, 1994; Popovich and Kaiser, 1995)). This alternative model operates more in the vein of a broadcast message server, such as Field (Reiss, 1995), where the only purpose of the Foundation is to forward notification messages that a particular SubEnv has *already* performed a particular task. This could be augmented with limited process support, as in Forest (Garlan and Ilias, 1990) and Provence (Krishnamurthy and Barghouti, 1993), to transform notification messages into requests to perform various tasks triggered by process enactment in the Foundation.

- **Data translation and transfer.** When process enactment at one SubEnv involves access to data "owned" by one or more other SubEnvs, the Foundation must provide mechanisms for transferring product artifacts and requisite process state among SubEnvs. As in homogeneous federations, bulk data may be temporarily cached, permanently copied, or migrated between sites. Note that enactment of such tasks may not be frequent in a multi-PCE, e.g., data exchange may be limited to scheduled milestones, whereas collaborative tasks are expected to be more commonplace in a multi-site PCE.

  A homogeneous federation can assume a standard data repository, although perhaps with differing local schemas, whereas heterogeneous federations also incur the problems of incompatible data formats; this is basically a distributed computing issue attacked by OMG (Object Management Group) and others through CORBA and similar layers, and not addressed further in this paper. Further, homogeneous federations assume compatible transaction management (concurrency control and failure recovery), generally supporting a two-phase commit protocol for distributed transactions, which may not be straightforward in the heterogeneous case. This issue has been addressed extensively in the database community, (e.g., Chen et al., 1993), and is also not discussed further here.
- **Federation configuration.** Finally, there should be some means for configuring a federation. We anticipate this is considerably more difficult and heavyweight for heterogeneous than for homogeneous federations, and in the former case may involve substantial design and implementation to introduce a new PCE (i.e., if it was not previously integrated with the Foundation) rather than just invoking a pre-defined (re)configuration process. Substantial effort may be involved in introducing the conceptually homogeneous infrastructure component mentioned above into a given PCE, and the mechanism for doing so is necessarily ad hoc (i.e., PCE-specific).

2.6. *Heterogeneous federation architectures*

The three main categories are outlined below.

2.6.1. Ad hoc. A handcrafted federation consisting of a very small number of distinct PCEs, e.g., Bull's ACME (Arnold and Popovich, 1994) integration of Conversation Builder

(Kaplan et al., 1992) and MARVEL 3.x. While one might be able to find a special-purpose Foundation component in this or similar examples, we are concerned in this paper with general federation. There is no distinction between the ad hoc approach and peer-to-peer architectures in the heterogeneous case, because by definition there is no common means for introducing a tightly coupled Foundation moderator, or equivalent code inside the SubEnv's internal architecture, nor any general way for a moderating component to incorporate SubEnv-specific knowledge of process and other concerns.

2.6.2. Centralized glue. The SubEnvs communicate/interact through a centralized component that implements the Foundation—the same architecture as shown in figure 2, except that the SubEnvs may be different (originally) single-site PCEs rather than components of the same multi-site PCE. This approach is exemplified by ProcessWall. Mentor takes a similar task, except that there may be *multiple* state servers and task servers, not just one (of each).

2.6.3. Decentralized glue. The Foundation is divided into multiple distributed components, i.e., intermediaries analogous to those attached to each SubEnv as in figure 3 and loosely coupled with their local SubEnv, except that each of the SubEnvs may be a different system. There is *no* centralized component.

2.7. *Choice of heterogeneous architecture*

When a wide range of internal architectures is exhibited among the PCEs of interest, there is usually no obvious preference exhibited for centralized vs. decentralized glue, except as in any distributed system a decentralized approach will generally scale better; however, the extra software engineering effort of separately retrofitting a large number of existing local PCEs is more likely to be the limiting factor than a central bottleneck. Thus the number of local PCEs is expected to be small. By analogy to the discussion of Section 2.4, when scaling is not an issue, top-down global processes would generally be more amenable to a centralized Foundation, and bottom-up to a decentralized Foundation. But the case is not so compelling for heterogeneous as for homogeneous federation, there considering glue vs. peer-to-peer, since construction of a global process using diverse process modeling languages and paradigms is so complex as to overwhelm all other concerns.

This may be why Heimbigner proposes a third model for his process task server, a hybrid of top-down and bottom-up: *constructor* PCEs post pending tasks to the shared process repository in a generally bottom-up fashion, whereas it makes more sense for the *constrainer* PCEs, which remove disallowed tasks from among those posted, to be organized top-down. That is, constructors create newly instantiated tasks according to local process workflow, but constrainers may disallow some tasks according to global process constraints. A distinguished constructor, effectively part of the Foundation, might post multi-site tasks implementing a top-down global process.

The discussion in Section 4 considers a heterogeneous federation architecture where OZ plays the role of a constructor (and employs its own constraints prior to instantiating a task to post); other PCEs integrated into the same federation via ProcessWall could, of course, act as constrainers on the posted tasks as well as additional constructors.

3. The Oz homogeneous PCE federation

OZ is the only fully implemented homogeneous PCE federation that we know of.[2] OZ originally (versions 1.1.1 and earlier) followed the direct peer-to-peer architectural model, where the majority of the Foundation functionality was built into the process engine (as elaborated in (Ben-Shaul and Kaiser, 1995b)). OZ was later reimplemented (versions 1.2 and later), using a new process engine as moderated peer-to-peer—with the Foundation moderator separated out into a component invoked via generic "callbacks" from the process engine. We are primarily concerned with the later form of OZ in this paper.

The overall choice of architecture for OZ follows the analysis given in Section 2.4. First, one of the major requirements for the OZ system is to support interoperability among autonomous, geographically distributed, and possibly pre-existing processes, e.g., the latter might have been designed for a single-site MARVEL environment; we developed a utility that mechanically upgrades an instantiated MARVEL environment to an OZ SubEnv. This requirement implies a bottom-up approach, which in turn suggests a peer-based architectural style. However, we have constructed both bottom-up (e.g., see (Ben-Shaul and Kaiser, 1995b)), Appendix A) and top-down (e.g., see (Kaiser et al., 1997)) global processes for OZ.

Second, OZ was developed (among other reasons) to interconnect instances of the MARVEL framework. Since MARVEL's client/server architecture corresponds to the centralized-process control and decentralized-tool execution local architecture, and MARVEL's process engine provided no API and the source code was handy, it was natural to adopt a direct peer-to-peer approach. Later on, OZ's native process engine adapted from MARVEL was replaced with the Amber process server (Popovich, 1997), which provides an API (and "callback" interface), and hardwires neither centralized-process control nor decentralized-tool execution. Amber itself was not modified at all to produce OZ multi-site functionality, as discussed in (Kaiser et al., 1996).

3.1. Marvel and Oz overview

Everything described here about MARVEL is also true for OZ unless stated otherwise.

MARVEL (Kaiser et al., 1988) provides a rule-based process modeling language in which a rule generally corresponds to a process step, or task. Each rule specifies the task's name as it would appear in a user menu or agenda; typed parameters and bindings of local variables from the project objectbase; a condition or prerequisite that must be satisfied before initiating the activity to be performed during the task; the tool script with in, inout, and out arguments for the activity (Gisi and Kaiser, 1991); and a set of effects, one of which asserts the actual results or consequences of completing the activity (some activities have more than one possible result). Built-in operations like `add`, `delete`, etc. are modeled as rules, and can be overloaded, e.g., to introduce type-specific conditions and effects on those operations; built-in operations can also be used as assertions in the effects of other rules.

MARVEL enforces that rule conditions are satisfied, and automates the process via forward and backward chaining. When a user requests to perform a task whose condition is not currently satisfied, the process engine backward chains to attempt to execute other rules whose effects may lead to satisfying the condition; if all possibilities are exhausted, the user is informed that the chosen task cannot be enacted at this time. When a rule completes, its asserted effect may trigger forward chaining to automatically enact other rules whose conditions have become satisfied. Users usually control the process by selecting rules representing entry points into composite tasks consisting of one main rule and a small number of auxiliary rules (reached via chaining) for change propagation and automation of menial chores, but it is possible to define complete workflows as a single goal-driven backward chain or event-driven forward chain.

MARVEL employs a client/server architecture (Ben-Shaul et al., 1993). Clients provide the user interface and execute tasks, usually by invoking external tools. The server context-switches among multiple clients, and includes the process engine, object management, and transaction management. OZ is essentially the same as MARVEL, except that an OZ environment may consist of several servers, each with its own distinct process model, data schema, objectbase and tools (Ben-Shaul and Kaiser, 1995b). Clients are always connected to one "local" server, and may also open and close connections to "remote" servers on demand. A server and its "local" clients constitute a SubEnv. The external view of the multi-site peer-to-peer OZ architecture is shown in figure 6.

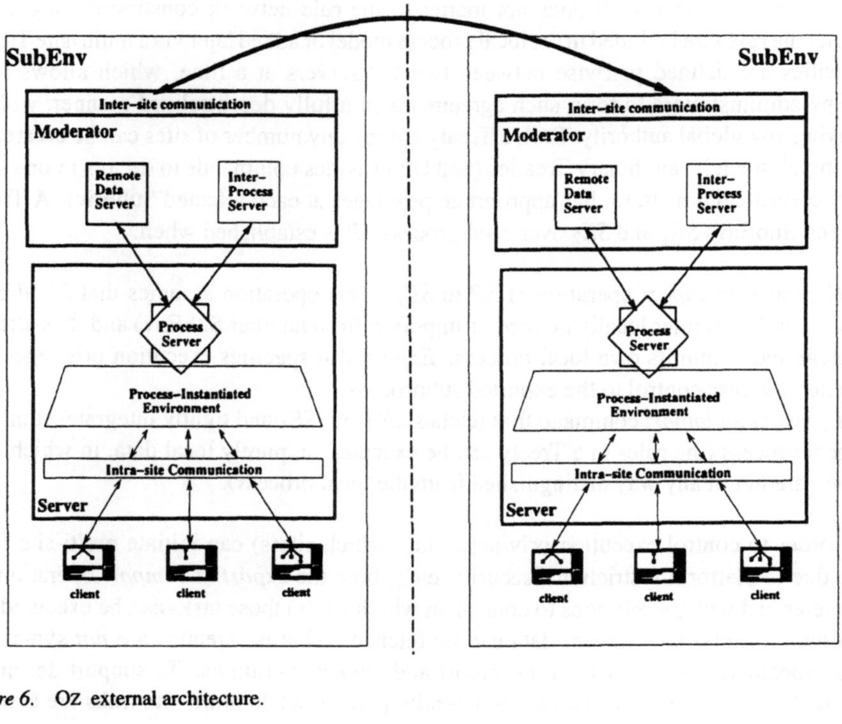

Figure 6. OZ external architecture.

OZ servers communicate with each other mainly to establish and operate *alliances*, which involves (1) negotiation of *Treaties*—dynamically agreed-upon shared subprocesses that are automatically and incrementally added on to each affected local process on the fly when the Treaty is instituted (and automatically and incrementally removed when a site unilaterally revokes the Treaty); and (2) coordination of *Summits*—enactment of Treaty-defined process segments that involve data and/or local clients from multiple sites, with computation interleaved between shared and local computational models (i.e., the Treaty and the local processes). We stretch the International Alliance metaphor, since Treaties among sites precede and specify Summits rather than vice versa.

3.1.1. Treaties and summits. The purpose of a Treaty is to establish a common subprocess. A Treaty consists primarily of a set of OZ rules. These rules define intentionally multi-site tasks, where the parameters are expected to be selected from multiple sites (i.e., distinct OZ objectbases chosen by a user via open connections to "remote" servers). Such tasks must be defined somewhere, in OZ's case within one of the participating SubEnvs. However, in general, Treaty rules are not an inherent component of any SubEnv's local process. Instead, the common subprocess is combined into each local participating process, in the sense that its tasks may be synchronized with other local tasks, depend on the outcome of their execution, and vice versa.

This is relatively easy to do with rules, the basis of OZ's process modeling formalism, since process enactment follows automatically determined forward and backward rule chains based on matching a predicate in one rule's condition to an assertion in another rule's effect (Heineman et al., 1992). It does not matter to the rule network construction algorithm whether the rules are included in the local process model or added later via a multi-site Treaty.

Treaties are defined pairwise between two OZ servers at a time, which allows local SubEnv administrators to form such agreements in a fully decentralized manner, without involving any global authority. Still, a Treaty among any number of sites can be created by forming all the relevant binary Treaties (and OZ provides commands to do this in one step, if the relevant administrator has appropriate privileges at each affected SubEnv). A Treaty between SubEnvs SE_1 and SE_2 over a subprocess SP is established when:

1. SE_1 issues an *export* operation of SP to SE_2. This operation assumes that SP already exists in SE_1 (either locally defined or imported from another SubEnv) and thus already integrated within its own local process. *Export* also specifies execution privileges and general access control to the exported subprocess.
2. SE_2 issues an *import* command that fetches SP from SE_1 and tightly integrates it into its local process (the rules in a Treaty can be executed on purely local data, in which case they are not in any way distinguished from the local process).

In order to control execution privileges, i.e., which site(s) can initiate multi-site tasks (e.g., due to platform restrictions, security, etc.), both the *export* and *import* operation are parameterized with permissions to control on which site(s) those tasks can be executed and from which site(s) the relevant data can be fetched. That is, Treaties are *not* symmetric unless specified as such by both the *export* and *import* operations. To support decentralization, Treaties may be withdrawn unilaterally (except while multi-site tasks are actually being executed); note this requires dynamic Treaty validation, i.e., checking that none of the affected parties to the Treaty have revoked it. Thus, not only does the Treaty mechanism allow definition of decentralized multi-site processes, but the (meta-) process for establishing and maintaining Treaties is itself highly decentralized.

Summits are the process enactment counterpart of Treaty process models. When a multi-site rule (from some Treaty) is issued for enactment by a user client of its "local" OZ server, termed hereafter the *Summit coordinator*, that SubEnv performs the following main steps:

1. Verify that the corresponding Treaty is valid (i.e., it has not been retracted by one of the SubEnvs whose data was selected as parameters to the rule).
2. Evaluate the rule's condition to determine whether or not it is satisfied. This requires fetching all the parameters from their home sites and catching them locally.
3. If the condition is not satisfied, send to those participating SubEnvs whose data fail their condition predicates a request to issue *local pre-Summit* tasks, which involve local (hence private) process steps on local data with local tools determined via backward chaining from the requested multi-site rule.
4. Wait for all sites to return before continuing to the next phase. Note that if the original rule's condition is already completely satisfied, then the pre-Summit phase is null.
5. Execute the multi-site activity of the rule, usually but not necessarily involving data from multiple sites (it is possible that remote parameters appear only in the condition and/or an effect), and possibly multi-site tools (e.g., groupware).
6. Send to each participating SubEnv a request to update its own data affected by assertions of the rule's actual effect (determined according to the return code of the rule's activity).
7. Each such SubEnv issues corresponding *local post-Summit* tasks determining via forward chaining from the relevant assertions of the original rule's effect, again involving only local resources.
8. Wait for all affected sites to reply.
9. Enact further related Summits, if any, reached via forward chaining to other multi-site rules from the original Summit rule.

Thus, Summits alternate between execution of shared, global, and multi-site tasks, to execution of private, local and single-site tasks, and effectively enact multi-site processes with minimal interprocess dependencies beyond the explicitly defined shared subprocesses. Full details of Summits and Treaties are given in (Ben-Shaul and Kaiser, 1995a).

Treaties and Summits impose several requirements on the design of OZ's federated architecture. First, the tight process-level integration of an imported subprocess into the local process implies a strong coupling of the Foundation with the local PCE. The strict decentralization, even in the definition of the federated process (using Treaties), avoids the need for a global process state (except for the special configuration process, described in (Ben-Shaul and Kaiser, 1994a)), and further supports our choice of a peer-to-peer architecture according to the issues discussed in Section 2.4.

Finally, Summits do not require a global process controller, but do require functional extensions to local process engines to allow them to become Summit coordinators. Again, the peer-to-peer architecture is favored. However, none of the above aspects indicate any preference with respect to direct vs. moderated (peer-to-peer) approaches, assuming the process engine can be extended into a coordinator without internal code modifications—as is the case for the Amber process server but which would have been difficult for Oz's native process engine. Thus this decision is likely to be based on the lower-level architectural and implementation aspects of the local SubEnv—and, as noted above, we have tried both models and prefer the moderated approach on modularity and extensibility grounds.

Note that none of these concerns are specific to rules, as opposed to some other process paradigm, except in the sense that the rule network generation algorithm makes it trivial to tightly bind imported rules with the local process. Integration of imported Treaties into the local process for non-rule process paradigms is more complicated, but possible, as discussed in (Ben-Shaul and Kaiser, 1995a) and not addressed here.

3.2. Oz architecture

The internal architecture of Oz is shown in figure 7. We use the following graphical notations: squared boxes with the widest bold lines (e.g., the Server) represent operating system processes, or independent threads of control; squared boxes with lines with

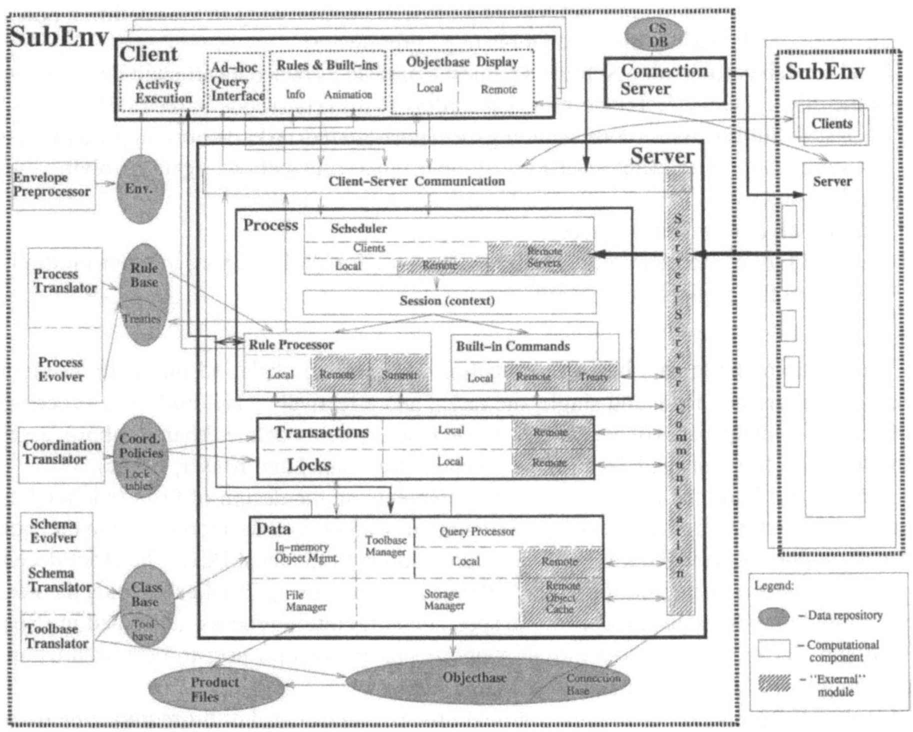

Figure 7. Oz internal architecture.

intermediate width (e.g., the Process component) represent top-level computational components that are part of the same operating system process as other components but are relatively independent from those components; squared boxes with narrow solid lines are computational subcomponents; dashed-line separators within subcomponents further modularize a (sub)component into its various functionalities; shaded rectangles within the above indicate "external" modules that extend the functionality of the basic component (as explained below); shaded ovals represent data repositories; and arrows represent data and/or control flow. The relative sizes of the various units are **not** intended to be meaningful.

Oz consists of three main runtime computational entities: the Environment Server (or simply, the *Server*), the *Connection Server*, and the *Client*. In addition, there are several utilities that convert the various project-specific definitions into an internal format that is understood and loaded by the server; they are of no concern in this paper.

There are three kinds of interconnections: client-to-local-server, client-to-remote-server, and server-to-server. The first connection is "permanent", in the sense that its existence is essential for the operation of the client. That is, a client is assumed to always be connected to its local server, and when such a connection becomes disconnected (either voluntarily on demand or involuntarily due to some failure) the client normally shuts down and is removed from the local server's state[3]. In contrast, the two other connections can be regarded as "temporary", since they are optional, and can be dynamically reconnected and disconnected over the course of a user session without disrupting the local operation of a SubEnv. This is a necessary feature to fulfill the independent-operation requirement, particularly when the servers are spread arbitrarily over multiple administrative domains.

An Oz (multi-site) environment consists of a set of instantiated SubEnvs, and at any point in time none, some, or all SubEnvs may be *active*. A SubEnv is considered active if exactly one server is executing "on the environment", meaning that it has loaded the SubEnv's process, and the SubEnv's objectbase (containing persistent product data and process state) is under the control of the server's data management subsystem (described in Programming Systems Laboratory (1996)). Typically an active environment also has at least one active (i.e., executing) client connected to its server, because the server automatically shuts itself down when there are no more active clients (and is automatically started up on demand by the Connection Server, as will be explained shortly).

3.2.1. The environment server. The server consists of three major components: process, transaction and data managers, each of which can be separately tailored by a combination of two facilities: declarative definitions loaded from a file and "external" code modules. The process manager loads the process model (including portions obtained through Treaty *import*), the transaction manager is parameterized by lock tables and concurrency control policies, and the data manager loads the schema for the product data and process state (currently imported rule sets must employ subschemas compatible with the local schema, although some conversion is supported). All of these tailorings are stored in environment-specific files; see Programming Systems Laboratory (1996) for details. The conceptually "external" code is hardwired into Oz's data manager[4], reasonably independent and invoked through a callback interface in the case of the process manager, and completely independent and dynamically loaded for the transaction manager. We do not consider the distinction between "external" vs. intrinsic code further in this paper, that is the subject of other papers ((Kaiser et al., 1996; Heineman and Kaiser, 1995), respectively).

Process manager. The process manager is the main component of the server. Its frontend subcomponent is the scheduler, which receives requests for service from three entities that correspond to the previously mentioned interconnections, namely local clients, remote clients, and remote servers. With few exceptions, notably to prevent deadlocks among mutual server communications, these requests are served on a first-come-first-served basis. The server is non-preemptive, i.e., it relinquishes control and context-switches only voluntarily.

The session layer encloses each interaction with a server in a context containing information that enables it to switch between and restore contexts. The context of locally executing tasks, including those that execute as part of a pre- or post-Summit, and the context of composite (multi-task) Summits, are represented in task data structures.

The rule processor consists of subcomponents for processing local tasks, local tasks spawned via pre-Summit or post-Summit processing from either local or remote Summits (denoted "Remote" in the figure), and Summit tasks. There are very few "system" built-in activities (notably parts of the configuration process), so the behavior of a particular instantiated SubEnv is mostly determined by the rule set that defines the process.

The built-in command processor handles all the kernel services that are available to every SubEnv. These include the primitive structural operations on the objectbase (e.g., `add` and `copy` object), several display options and image refresh commands, access control, and the various dynamic process loading and Treaty operations.

In the original direct peer-to-peer variant of OZ, all alliance support was hardwired. But in the newer moderated peer-to-peer versions, Summits, Treaties and related infrastructure has been culled out into the "external" code modules indicated in the figure.

Transaction manager. All access to data is mediated by OZ's transaction manager. Due to the required decentralization, each transaction manager is inherently *local*, i.e., it is responsible only for its local objectbase. However, transaction managers attached to each server communicate among themselves to support concurrency control and failure recovery involving remote objects. OZ's transaction manager was developed separately and has been used independent from the rest of OZ. Further details are outside the scope of this paper, see (Heineman, 1996).

Data manager. This component includes an in-memory object manager that provides uniform object-based access to data from any system component. Objects can be looked up in one of three ways: by structural navigation, by class membership, and by their object-identifier (OID). Structural and by-class searches are requested by the query processor to service navigational and associative queries, respectively, and by-OID lookup is used for several purposes, among them to support direct user selection of objects (mouse clicking in the objectbase display) as parameters to rules.

The second major subcomponent is the query processor. It supports a declarative query language interface, and is called from both the rule processor for embedded queries and directly from the user client for servicing ad hoc queries. Queries on remote objects are handled at this level, by invoking a server-to-server service.

The rest of data management consists of an untyped storage manager (implemented on top of the `gdbm` package) that stores the objectbase contents; a file manager that handles access to file attributes (file attributes are paths to files resident in the environment's "hidden file system"); and an object cache that holds transient copies of remote objects during Summits.

The data manager is tailored by the project-specific schema tied to the instantiated objectbase, including both class- and composition-hierarchies. As in the case of rules and the process manager, without a schema the data manager is useless since it cannot instantiate any objects.

3.2.2. The client. There are three main clients, supporting X View, Motif and tty (command line) user interfaces, as well as several auxiliary clients with no user interface intended primarily for tool execution. Each client consists of four major subcomponents: (1) access to information about rules and built-in commands, (2) objectbase representation, (3) activity execution, and (4) an ad hoc query interface.

The two graphical user interface clients are (conceptually) multi-threaded, i.e., a single client can support multiple concurrent interactions with local or remote servers. This enables a user to run in parallel several (possibly long) activities from the same client. The command interface includes a process-specific menu and utilities for displaying rule definitions and the rule network interconnections, all of which are stored in the client's address space and can be dynamically refreshed when a new process is (re)loaded or a Treaty is formed. A dynamic rule-chaining animator shows the control flow of enacted tasks as they execute, both local and Summits.

The objectbase display maintains an "image" of *structural* information, i.e., parent/child and reference relationships, for browsing and for selecting arguments to activities. The contents of primitive and file attributes are transmitted only when needed. Users can select the `open-remote` command to display the objectbase images from other sites and subsequently select objects from multiple sites, allowing invocation of a Summit rule. The client maintains multiple simultaneous connections to the remote servers, and is able to direct requests to appropriate servers.

3.2.3. The connection server. The Connection Server's main responsibility is to (re)establish connections to a local server from local clients, remote clients, and remote servers. However, it does not participate in the actual interactions between those entities; it serves only as a mediator for "handshaking" purposes. In some cases, the destination server to which a request for a connection is made may not be active, in which case the Connection Server is capable of automatically (re)activating the dormant server. In other cases the desired server may be active but its address (host IP address and port number) might be unknown to the requesting entity, in which case the Connection Server sends that information to the requesting entity for further communication.

Unlike the Environment Server, the Connection Server is (conceptually) always active, since it is implemented as a daemon invokable via the Unix `inetd` mechanism. Thus, each configured host has its own (logical) Connection Server that supports all SubEnvs (of the same or different multi-site environments) that reside on that host.

4. Oz/ProcessWall: A hypothetical heterogeneous federation

A heterogeneous federation is inherently more general than a homogeneous federation. Thus it is desirable to consider how a multi-site PCE like Oz might "fit" into a multi-PCE organized via a process state/task server, the only specific model we know is for heterogeneous federation of PCEs. Note the federation would presumably also include various non-Oz SubEnvs.

One approach is to drop the homogeneous Foundation entirely and employ only the heterogeneous Foundation for multi-site tasks. Then the homogeneous SubEnvs—i.e., homogeneous with respect to system but heterogeneous with respect to process model—would be treated as if they were unrelated local PCEs rather than part of an Oz multi-site PCE. Assuming that some component is added to interface with the federation glue, this should work trivially if they fulfill the requirement of independent operation—that is, that they do not depend on each other in any way to perform entirely local work. In other words, in principle, we could have used ProcessWall to scale up MARVEL to support process interoperability across multiple teams, each with their own local MARVEL environment instance. But then the main advantage of a homogeneous federation is lost, namely the relative ease with which SubEnvs can call on each other to perform specific agreed-upon services within the identical (and thus mutually understood) process modeling and enactment paradigm.

An alternative approach is to allow individual (or all) SubEnvs of a homogeneous federation to participate in one or more heterogeneous federations, while retaining the higher level of intimacy afforded by the system-level homogeneity when (intentionally) interacting with other local components of the same system.

Note that SubEnvs that happen to participate in the same homogeneous federation may happen to employ each other's services indirectly through the heterogeneous federation, without necessarily any knowledge that they have more direct means of interaction. In fact, through this "backdoor" one might, in an unusual circumstance, inadvertently arrive in a situation where a SubEnv indirectly requests services from itself without realizing that its doing so—which could potentially happen in a homogeneous federation as well, although not in the Oz realization because the server checks for this degenerate case.

4.1. *Issues*

Assuming a centralized Foundation in the style of the ProcessWall state/task server, the main questions to answer are:

1. How would an Oz SubEnv post to the Foundation those tasks it has instantiated but not initiated, and (generally speaking) would like some other PCE to perform?
   The local task descriptor must of course be converted to the Foundation's standard form. There are complications regarding representation of data arguments as part of the task specification, and later regarding data transfer when the task is enacted by some PCE participating in the federation. Even though the task may eventually be picked up by another Oz SubEnv, this cannot be assumed a priori (if it could, direct interaction through the homogeneous Foundation would almost certainly be more efficient).

2. How would the Foundation notify an Oz SubEnv of the completion of such a task?
The Foundation's relevant state must be converted to a form understood by Oz, and either automatically transmitted to Oz by the Foundation or explicitly requested by Oz, e.g., via periodic polling or some kind of rendezvous. When data arguments are modified during the task, changed data must either be submitted back to the Oz objectbase, or Oz must be notified of its whereabouts and have some means to retrieve the data.
3. How would the Foundation inform a particular Oz SubEnv that it should perform a specific posted task?
One approach involves some kind of scheduler or other entity that selects among enabled tasks for enactment, chooses the recipient SubEnv, and sends an appropriate request to that SubEnv. This Foundation-generated request model is compatible with Oz's current server-to-server communication mechanism. In contrast, a completely new interface would be needed to fit into a blackboard model that required each SubEnv to poll the Foundation for suitable enabled tasks. However, a hybrid might be achieved in the style of a broadcast message server like Field, where the SubEnv's *register* their interests in or abilities to perform certain kinds of tasks, perhaps by supplying a pattern that is matched by the Foundation against the enabled task specifications. The application of event subscription to workflow management system interoperability is suggested in (Schwenkreis, 1996).
4. How would an Oz SubEnv notify the Foundation of a task it had just completed?
There are two cases: the task was previously posted to the Foundation by the same or a different SubEnv, or it arose entirely inside the given SubEnv (and thus is supplied only as historic information). In the former case, the Foundation might have requested that this SubEnv perform the task, or alternatively the SubEnv might have selected the task from among those enabled (via either polling or registration). Note that generally it is necessary for the Foundation to prevent multiple SubEnvs from concurrently agreeing to perform the same posted task. Again, data transfer is of concern here in both directions.

4.2. Integrating Oz *tasks and processwall tasks*

There are three different "levels" of task-like units supported by Oz: rule activities, full rules, and rule chains, any or all of which could be mapped to ProcessWall's notion of tasks. The answers to the questions above will be somewhat different depending on which Oz unit is chosen for the mapping.

Oz's lowest task-like "level" is an individual rule activity, i.e., invocation of a tool script (and thence external tool). Oz already defines a client/server protocol whereby user interface clients tell the server to apply a selected rule to a list of objects and literal arguments; once the condition is deemed satisfied, the server supplies the client with corresponding file pathname and primitive arguments, and directs it to invoke the tool script specified in the rule activity; the client forks the tool script, and the tool script or the tool(s) it invokes are assumed to directly modify the contents of file arguments; finally, the client returns to the server with the return code from the tool script, which selects among the possible rule effects, as well as (optional) assignments to output variables. The encapsulating rule (and its pending chain) then continues.

It is simple to construct a special Oz client that receives the same message from the server identifying tool script and arguments but does something different than the typical user client; in fact, we have already introduced numerous such clients (see (Skopp, 1995; Dossick and Kaiser, 1996; Valetto and Kaiser, 1996; Lee et al., 1996)). Then, to implement points 1 and 2, the new client would be inserted into the multi-PCE architecture between the Oz server and the central Foundation. This client would convert the activity information provided into the Foundation's task representation and forward it to the centralized Foundation. Later, after the activity has been completed, the Foundation would notify this special client, which would then respond to the Oz server like any of its other clients. The special client would be responsible for data traffic in both directions.

The same special Oz client (or alternatively a distinct special client) can be used to implement points 3 and 4. The Foundation sends the task to the Oz client for that client to execute itself using the same tool invocation facilities as any other Oz client, without involving the Oz server. When the task completes, the special Oz client returns the results to the Foundation, again without interaction with the Oz server. Note that some special communication facilities will be needed in the Oz client, if the same client is used for both purposes, to avoid deadlock when the client happens to be forwarding an activity to the Foundation to be performed by some other PCE at the same time that the Foundation is sending a request to the client.

The intermediate task "level" corresponds to entire Oz rules, with condition and effect(s) as well as activity. Oz servers already transmit rule definitions between themselves as part of Treaty negotiation, and transmit the parameters and bound variables of instantiated rules as part of Summit enactment. The newer variants of Oz introduce a protocol for transferring instantiated rules between client and server to support delegation to and selection from user and group agendas ("to do" lists) (Kaiser et al., 1997). Rules with and without already satisfied conditions may appear in an agenda. These facilities might be combined and extended, again through a translating client interposed between Oz and the Foundation, to support all four points.

A complication: In the lowest level case the condition or prerequisite is already satisfied, by definition, prior to posting the activity, but this would not in general be true in the intermediate case. Analogously, handling the effect or consequence of the activity is the concern of only the originally posting PCE, but again this cannot be assumed in the intermediate case. The ProcessWall and Mentor task representations allow for predecessors and successors, but not all the constraints embodied in Oz conditions are concerned with checking simple predecessor relationships (e.g., the local variable bindings might find all objects that match a complex associative query and then the condition checks that at least one of those objects satisfies a complex logical clause), nor are all assertions made in Oz effects concerned with triggering successors (e.g., objects can be created and deleted, reference links formed and removed, etc. through invocation of built-in operations).

One could argue for a simplification, whereby Oz's postings to the Foundation are limited to those tasks whose conditions and effects are solely concerned with predecessor/successor relations that can be directly represented by the Foundation's state and/or task model. Although Oz's process modeling language tends to obscure such relationships from a human-readability standpoint, they are visible in the internal rule network compiled from the process model. Since the Foundation can only represent such relations, by definition any request sent by the Foundation to Oz would so restrict the implicit conditions and effects.

A better approach might be to extend the Foundation's task representation, or develop some additional control channel, for transmitting the conditions and/or effects from the Oz SubEnv to the (potentially) foreign SubEnv for evaluation within its paradigm, and vice versa regarding communicating any prerequisites and consequences that might be supported by the foreign paradigm to an Oz SubEnv (and of course both issues come up between pairs of non-Oz SubEnvs as well). If Oz were configured as a multi-site homogeneous federation where some (or all) sites happened to also belong to a heterogeneous federation, pending tasks posted through the Foundation to another Oz SubEnv (in the same federation) could include their conditions and effects in some *opaque* data stream understood only by Oz servers.

So difficulties arise only when pending tasks posted through the Foundation involve non-Oz SubEnvs. Fortuitously, we have already shown fairly straightforward mappings from most of the major PCE paradigms, including Petri nets (Popovich and Kaiser, 1995), task graphs (Heineman, 1994), and grammars (Kaiser et al., 1994), into Oz rules, and reverse mappings are not inconceivable. As previously noted in Section 1, Mentor supports translation from one notation into another, as does the process interchange format standardization effort.

However, the general case requires substantial translation capabilities regarding data formats, and predicates and operations over those formats. The "universal data model" problem is a well-known unresolved, probably unresolvable issue in database research (Neuhold and Stonebraker, 1989). It may be possible to address a special case of this problem with respect to PCEs, e.g., if we assume the main data arguments are files and all attributes that might be referred to regarding task prerequisites and consequences (rule conditions and effects in Oz) are standard file system appendages supported by most operating systems, such as owner, read/write timestamps, access permissions, etc., or encoded predecessor/successor relationships modeled directly by the Foundation.

The third task "level" in Oz is a rule chain, i.e., all the rules emanating from some user-selected (or Foundation-requested) rule through backward and/or forward chaining. This seems easiest to handle by iterating the intermediate case as the rule chain unfolds, in the Oz to Foundation case, or incrementally sending each member of a sequence of predecessor/successor tasks, in the Foundation to Oz case.

Note, we have so far ignored the issue of specifying which PCE *user* should perform the work in the case of an interactive task. Oz version 1.1.1 included process modeling and enactment facilities, which could be revived, to delegate a rule activity to a specific user, or to one or all members of a user group (Ben-Shaul and Kaiser, 1996). Later versions of Oz can delegate an entire rule or (rest of a) rule chain to a specific user or any group member via "guidance chaining", a form of forward chaining where the next rule in the chain is placed in an agenda rather than immediately enacted (Tong et al., 1994). How *another* PCE might designate a user to perform tasks originally instantiated by Oz is of course open-ended.

5. Contributions and future directions

The main contributions of this work are:

- The elaboration of requirements and architectures for homogeneous and heterogeneous federations of process-centered environments. Both the homogeneous and heterogeneous federation architectures we present are in line with a proposed distributed workflow reference model (Wallnau et al., 1996).
- The design and realization of a specific homogeneous federation architecture for Oz.
- A presentation of the issues that must be addressed to integrate Oz into a heterogeneous federation based on the ProcessWall process state/task server (or Mentor worklist/history manager) approach.

The obvious next step is to complete an experimental integration between Oz and the realization of a process state/task server, assuming one becomes available. It would be desirable to also include in the heterogeneous federation at least one other PCE, besides Oz. Evaluation against the heterogeneous federation requirements should prove interesting.

Finally, we would like to introduce greater flexibility into Oz alliances, i.e., homogeneous federations, including navigation and search among related SubEnvs both within and across alliances, easy movement of user clients from one SubEnv server to another, and lighter weight composition and destruction of SubEnv alliances. Note the second point presumes support for arbitrary geographical dispersion *within* a SubEnv, not just *among* SubEnvs. Thus, Oz user clients could not continue to assume a shared network file system for accessing objectbase file attributes and communication bandwidth may become a concern, issues already addressed to a limited extent for our Oz "low bandwidth clients" in (Skopp, 1995).

Acknowledgments

We would like to thank Steve Dossick, George Heineman, Wenyu Jiang, Wenke Lee, Steve Linde, Steve Popovich, Peter Skopp, Jack Jingshuang Yang and Sonny Xi Ye for their various work on Oz, Amber and the replacement of Oz's native process engine by Amber. We also thank Dennis Heimbigner, Alex Wolf and George Heineman for their first cut towards a hypothetical Oz/ProcessWall integration as part of a funding agency presentation. Finally, we thank the anonymous reviewers for detailed and useful criticisms, comments and corrections. Oz is available for licensing at no cost for research and educational purposes; send email to MarvelUS@cs.columbia.edu for information. Connect to http://www.psl.cs.columbia.edu/for up-to-date information about Oz and related projects.

Notes

1. Some authors have used the terms "multi-site" and "geographically distributed" interchangeably, but here they refer to different concepts.

2. The Programming Systems Lab used a multi-site Oz environment to support all our day-to-day software development for about two years, since April 1995, although as of this writing we are transitioning to the OzWeb extension of Oz, which operates over the World Wide Web and supports a hypercode representation of product artifacts (Kaiser et al., 1997).
3. An extension of this model, in which clients can be disconnected from their server and continue to operate independently to enact a process segment until reconnection, has been investigated separately to support mobile computing (Skopp and Kaiser, 1993).
4. Such code has been separated out in the later OzWeb.

References

Alonso, G., Agrawal, D., El Abbadi, A., Mohan, C., Gunthor, R., and Kamath, M.U. 1995. Exotica/FMQM: A persistent message-based architecture for distributed workflow management. *IFIP WG 8.1 Workgroup Conference on Information Systems Development for Decentralized Organizations*, Trondheim, Norway.

Arnold, J.E. and Popovich, S.S. 1994. Integrating, customizing and extending environments with a message-based architecture. Technical Report CUCS-008-95, Columbia University, Department of Computer Science. The research described in this report was conducted at Bull HN Information Systems, Inc. ftp://ftp.psl.cs.columbia.edu/pub/psl/CUCS-008-95.ps.Z.

Baentsch, M., Molter, G., and Sturm, P. 1995. WebMake: Integrating distributed software development in a structure-enhanced Web. *3rd International World-Wide Web Conference*, Darmstadt, Germany. Elsevier Science B.V., http://www.igd.fhg.de/www/www95/proceedings/papers/-51/WebMake/WebMake.html.

Bandinelli, S. and Fuggetta, A. 1993. Computational reflection in software process modeling: The SLANG approach. *15th International Conference on Software Engineering*, Baltimore MD: IEEE Computer Society Press, pp. 144–154.

Barghouti, N.S. 1992. Supporting cooperation in the MARVEL process-centered SDE. In H. Weber (Ed.), *5th ACM SIGSOFT Symposium on Software Development Environments*, Tyson's Corner VA, pp. 21–31, Special issue of *Software Engineering Notes*, 17(5) ftp://ftp.psl.cs.columbia.edu/pub/psl/sde92.ps.Z.

Barrett, D.J., Clarke, L.A., Tarr, P.L., and Wise, A.E. 1996. A framework for event-based software integration. *ACM Transactions on Software Engineering and Methodology*, 5(4):378–421.

Ben-Shaul, I.Z., Kaiser, G.E., and Heineman, G.T. 1993. An architecture for multi-user software development environments. *Computing Systems, The Journal of the USENIX Association*, 6(2):65–103. ftp://ftp.psl.cs.columbia.edu/pub/psl/CUCS-012-92.ps.Z.

Ben-Shaul, I.Z. and Kaiser, G.E. 1994a. A configuration process for a distributed software development environment. *2nd International Workshop on Configurable Distributed Systems*, Pittsburgh PA, pp. 123–134. ftp://ftp.psl.cs.columbia.edu/pub/psl/CUCS-022-93.ps.Z.

Ben-Shaul, I.Z. and Kaiser, G.E. 1994b. A paradigm for decentralized process modeling and its realization in the Oz environment. *16th International Conference on Software Engineering*, Sorrento, Italy. IEEE Computer Society Press, pp. 179–188. ftp://ftp.psl.cs.columbia.edu/pub/psl/CUCS-024-93.ps.Z.

Ben-Shaul, I.Z. and Kaiser, G.E. 1995a. An interoperability model for process-centered software engineering environments and its implementation in Oz. Technical Report CUCS-034-95, Columbia University Department of Computer Science. ftp://ftp.psl.cs.columbia.edu/pub/psl/CUCS-034-95.ps.Z.

Ben-Shaul, I. and Kaiser, G.E. 1995b. *A Paradigm for Decentralized Process Modeling*. Boston: Kluwer Academic Publishers.

Ben-Shaul, I.Z. and Heineman, G.T. 1996. A three-level atomicity model for decentralized workflow management systems. *Distributed Systems Engineering Journal*, 3(4):239–249.

Ben-Shaul, I.Z. and Kaiser, G.E. 1996. Integrating groupware activities into workflow management systems. *7th Israeli Conference on Computer Systems and Software Engineering*, Herzliya, Israel. IEEE Computer Society Press, pp. 140–149. ftp://ftp.psl.cs.columbia.edu/pub/psl/CUCS-002-95.ps.Z.

Boehm, B. (Ed.) 1996. *10th International Software Process Workshop: Process Support of Software Product Lines*, Ventron, France.

Bolcer, G.A. and Taylor, R.N. 1996. Endeavors: A process system integration infrastructure. In Wilhelm Schäfer (Ed.), *4th International Conference on the Software Process: Software Process—Improvement and Practice*, Brighton, UK. pp. 76–89. http://www.ics.uci.edu/pub/endeavors.

Chen, J., Bukhres, O.A., and Elmagarmid, A.K. 1993. IPL: A multidatabase transaction specification language. *13th International Conference on Distributed Computing Systems*, Pittsburgh PA: IEEE Computer Society Press, pp. 439–448.

Conradi, R., Osjord, E., Westby, P.H., and Liu, C. 1991. Initial software process management in EPOS. *Software Engineering Journal*, 6(5):275–284.

Cusumano, M. and Selby, R.W. 1995. *Microsoft Secrets: How the World's Most Powerful Software Company Creates Technology, Shapes Markets, and Manages People*. New York: The Free Press.

Dossick, S.E. and Kaiser, G.E. 1996. WWW access to legacy client/server applications. *5th International World Wide Web Conference*, Paris, France. Elsevier Science B.V. pp. 931–940. Special issue of *Computer Networks and ISDN Systems, The International Journal of Computer and Telecommunications Networking*, 28(7-11). http://www.psl.cs.columbia.edu/papers/CUCS-003-96.html.

Fernström, C. 1993. Process WEAVER: Adding process support to UNIX. *2nd International Conference on the Software Process: Continuous Software Process Improvement*, Berlin: Germany. IEEE Computer Society Press, pp. 12–26.

Garlan, D. and Ilias, E. 1990. Low-cost, adaptable tool integration policies for integrated environments. In R.N. Taylor (Ed.), *4th ACM SIGSOFT Symposium on Software Development Environments*, Irvine CA, pp. 1–10. Special issue of *Software Engineering Notes*, 15(6).

Gisi, M.A. and Kaiser, G.E. 1991. Extending a tool integration language. In M. Dowson (Ed.), *1st International Conference on the Software Process: Manufacturing Complex Systems*, Redondo Beach CA: IEEE Computer Society Press, pp. 218–227. ftp://ftp.psl.cs.columbia.edu/pub/psl/icsp91.ps.Z.

Hayes-Roth, B. 1985. A blackboard architecture for control. *Artificial Intelligence*, 26:251–321.

Heimbigner, D. 1990. Proscription versus prescription in process-centered environments. In T. Katayama (Ed.), *6th International Software Process Workshop: Support for the Software Process*, Hakodate: Japan. IEEE Computer Society Press, pp. 99–102.

Heimbigner, D. 1992. The ProcessWall: A process state server approach to process programming. In H. Weber (Ed.), *5th ACM SIGSOFT Symposium on Software Development Environments*, Tyson's Corner VA, pp. 159–168. Special issue of *Software Engineering Notes*, 17(5).

Heineman, G.T. 1994. Automatic translation of process modeling formalisms. *1994 Centre for Advanced Studies Conference (CASCON)*, Toronto ON: Canada. IBM Canada Ltd. Laboratory, pp. 110–120. ftp://ftp.psl.cs.columbia.edu/pub/psl/CUCS-036-93.ps.Z.

Heineman, G.T. 1996. A Transaction Manager Component for Cooperative Transaction Models. Ph.D. Thesis, Columbia University Department of Computer Science, CUCS-010-96. ftp://ftp.psl.cs.columbia.edu/pub/psl/CUCS-010-96.ps.gz.

Heineman, G.T., Kaiser, G.E., Barghouti, N.S., and Ben-Shaul, I.Z. 1992. Rule chaining in MARVEL: Dynamic binding of parameters. *IEEE Expert*, 7(6):26–32. ftp://ftp.psl.cs.columbia.edu/pub/psl/expert92.ps.Z.

Heineman, G.T. and Kaiser, G.E. 1994. Integrating a transaction manager component with Process-WEAVER. Technical Report CUCS-012-94, Columbia University Department of Computer Science. ftp://ftp.psl.cs.columbia.edu/pub/psl/CUCS-012-94.ps.Z.

Heineman, G.T. and Kaiser, G.E. 1995. An architecture for integrating concurrency control into environment frameworks. *17th International Conference on Software Engineering*, Seattle WA: ACM Press, pp. 305–313. ftp://ftp.psl.cs.columbia.edu/pub/psl/CUCS-021-94.ps.Z.

Heineman, G.T. and Kaiser, G.E. 1997. The CORD approach to extensible concurrency control. *Thirteenth International Conference on Data Engineering*, Birmingham, UK. ftp://ftp.psl.cs.columbia.edu/pub/psl/CUCS-024-95.ps.gz.

Humphrey, W. and Kellner, M.I. 1989. Software process modeling: Principles of entity process models. *11th Internation Conference on Software Engineering*, Pittsburgh PA: IEEE Computer Society Press, pp. 331–342.

Jin, W., Ness, L., Rusinkiewicz, M., and Sheth, A. 1993. Concurrency control and recovery of multi-database work flows in telecommunication applications. *ACM SIGMOD International Conference on Management of Data*, Washington DC, pp. 456–459. Special issue of *SIGMOD Record*, 22:2.

Kaiser, G.E. and Feiler, P.H. 1987. An architecture for intelligent assistance in software development. *9th International Conference on Software Engineering*, Monterey CA: IEEE Computer Society Press, pp. 180–188. ftp://ftp.psl.cs.columbia.edu/pub/psl/icse87.ps.gz.

Kaiser, G.E., Feiler, P.H., and Popovich, S.S. 1988. Intelligent assistance for software development and maintenance. *IEEE Software*, 5(3):40–49. ftp://ftp.psl.cs.columbia.edu/pub/psl/CUCS-401-88.tar.Z.

Kaiser, G.E., Barghouti, N.S., and Sokolsky, M.H. 1990. Experience with process modeling in the MARVEL software development environment kernel. In B. Shriver (Ed.), *23rd Annual Hawaii International Conference on System Sciences*, Kona HI, Vol. II, pp. 131–140. ftp://ftp.psl.cs.columbia.edu/pub/psl/CUCS-446-89.ps.gz.

Kaiser, G.E., Popovich, S.S., and Ben-Shaul, I.Z. 1994. A bi-level language for software process modeling. In W.F. Tichy (Ed.), *Configuration Management*, in *Trends in Software*, John Wiley & Sons, chapter 2, no. 2 pp. 39–72. ftp://ftp.psl.cs.columbia.edu/pub/psl/CUCS-016-92.ps.Z.

Kaiser, G.E., Ben-Shaul, I.Z., Popovich, S.S., and Dossick, S.E. 1996. A metalinguistic approach to process enactment extensibility. In W. Schäfer (Ed.), *4th International Conference on the Software Process: Improvement and Practice*, Brighton, UK, pp. 90–101. ftp://ftp.psl.cs.columbia.edu/pub/psl/CUCS-016-96.ps.gz.

Kaiser, G.E., Dossick, S.E., Jiang, W., and Jingshuang Yang, J. 1997. An architecture for WWW-based hypercode environments. *1997 International Conference on Software Engineering: Pulling Together*, Boston: MA, pp. 3–13. ftp://ftp.psl.cs.columbia.edu/pub/psl/CUCS-037-96.ps.gz.

Kaiser, G.E, Heineman, G.T., Skopp, P.D., and Yang, J.J. 1997. Incremental process support for component-based software engineering. Technical Report CUCS-007-96, Columbia University Department of Computer Science. ftp://ftp.psl.cs.columbia.edu/pub/psl/CUCS-007-96.ps.gz.

Kaplan, S.M., Tolone, W.J., Carroll, A.M., Bogia, D.P., and Bignoli, C. 1992. Supporting collaborative software development with ConversationBuilder. In H. Weber (Ed.), *5th ACM SIGSOFT Symposium on Software Development Environments*, Tyson's Corner VA, pp. 11–20. Special issue of *Software Engineering Notes*, 17(5).

Korth, H.F. 1986. Extending the scope of relational languages. *IEEE Software*, 3(1):19–28.

Korth, H.F. 1995. The double life of the transaction abstraction: Fundamental principles and evolving system concepts. In U. Dayal, P. M. D. Gray, and S. Nishio (Eds.), *21st International Conference on Very Large Data Bases*, Zurich, Switzerland, pp. 2–6. Invited paper.

Krishnamurthy, B. and Barghouti, N.S. 1993. Provence: A process visualization and enactment environment. In I. Sommerville and M. Paul (Eds.), *4th European Conference on Software Engineering*, of *Lecture Notes in Computer Science*, Vol. 717 pp. 151–160. Springer-Verlag, Garmisch-Partenkirchen, Germany.

Lee, J., Yost, G., and the PIF Working Group. 1994. The PIF process interchange format and framework. http://www-sloan.mit.edu/ccs/pifmain.html.

Lee, W., Kaiser, G.E., Clayton, P.D., and Sherman, E.H. 1996. OzCare: A workflow automation system for care plans. In J.J. Cimino (Ed.), *1996 American Medical Informatics Association Annual Fall Symposium*, Washington DC, pp. 577–581. ftp://ftp.psl.cs.columbia.edu/pub/psl/CUCS-012-96.ps.Z.

Neuhold, E. and Stonebraker, M. (Eds.). 1989. Future directions in DBMS research. *SIGMOD Record*, 18(1):17–26.

Object Management Group. http://www.omg.org/.

Osterweil, L.J. 1995. *Presentation at Software Process Architectures Workshop*.

Peuschel, B. and Wolf, S. 1993. Architectural support for distributed process centered software development environments. In W. Schäfer (Ed.), *8th International Software Process Workshop: State of the Practice in Process Technology*, Wadern, Germany, pp. 126–128. Position paper.

Popovich, S.S. and Kaiser, G.E. 1995. Integrating an existing environment with a rule-based process server. Technical Report CUCS-004-95, Columbia University Department of Computer Science. ftp://ftp.psl.cs.columbia.edu/pub/psl/CUCS-004-95.ps.Z.

Programming Systems Lab. *Darkover Manual*. 1996. ftp://ftp.psl.cs.columbia.edu/pub/psl/oz.1.2.manuals/IV.Darkover_API/.

Programming Systems Laboratory. *Oz 1.2 Manual Set*. 1996. Columbia University Department of Computer Science. ftp://ftp.psl.cs.columbia.edu/pub/psl/oz.1.2.manuals.

Pu, C. 1989. Superdatabases for composition of heterogeneous databases. In A. Gupta (Ed.), *Integration of Information Systems: Bridging Heterogeneous Databases*, IEEE Press, pp. 150–157. Also appeared in *4th International Conference on Data Engineering*, 1988, Los Angeles CA.

Ram, S. (Ed.). 1991. Special Issue on Heterogeneous Distributed Database Systems, *Computer*. IEEE Computer Society Press, Vol. 24. p. 12.

Reiss, S.P. 1995. *The Field Programming Environment: A Friendly Integrated Environment for Learning and Development*. Boston: Kluwer Academic Publishers.

Schäfer, W. (Ed.). 1996. *4th International Conference on the Software Process*, Brighton: UK. IEEE Computer Society Press.

Schäfer, W., Peuschel, B., and Wolf, S. 1992. A knowledge-based software development environment supporting cooperative work. *International Journal on Software Engineering and Knowledge Engineering*, 2(1):79–106.

Schwenkreis, F. 1996. Workflow for the German federal government. *NSF Workshop on Workflow and Process Automation in Information Systems: State-of-the-Art and Future Directions*, Athens GA, pp. 64–68. Position paper.

Sheth, A.P. and Larson, J.A. 1990. Federated database systems for managing distributed, heterogeneous, and autonomous databases. *ACM Computing Surveys*, 22(3):183–236.

Sheth, A., Georgakopoulos, D., Joosten, S.M.M., Rusinkiewicz, M., Scacchi, W., Wileden, J., and Wolf, A. 1996. Report from the NSF workshop on workflow and process automation in information systems. Technical Report UGA-CS-TR-96-003, University of Georgia Department of Computer Science. http://lsdis.cs.uga.edu/activities/NSF-workflow/final-report-cover.html. /test.html.

Shy, I., Taylor, R., and Osterweil, L. 1989. A metaphor and a conceptual framework for software development environments. In F. Long (Ed.), *Software Engineering Environments International Workshop on Environments*, volume 467 of *Lecture Notes in Computer Science*, Chinon: France, Springer-Verlag, pp. 77–97.

Skopp, P.D. 1995. Low Bandwidth Operation in a Multi-User Software Development Environment. Master's Thesis, Columbia University, Department of Computer Science. CUCS-035-95. ftp://ftp.psl.cs.columbia.edu/pub/psl/CUCS-035-95.ps.Z.

Skopp, P.D. and Kaiser, G.E. 1993. Disconnected operation in a multi-user software development environment. In B. Bhargava (Ed.), *IEEE Workshop on Advances in Parallel and Distributed Systems*, Princeton NJ, pp. 146–151. ftp://ftp.psl.cs.columbia.edu/pub/psl/CUCS-026-93.ps.Z.

Sutton, S.M. Jr., Heimbigner, D., and Osterweil, L.J. 1995. APPL/A: A language for software process programming. *ACM Transactions on Software Engineering and Methodology*, 4(3):221–286.

The Common Object Request Broker: Architecture specification revision 2.0. 1996. Technical Report ptc/96-08-04, Object Management Group. http://www.omg.org/corba/corbiiop.htm.

Popovich, S.S. 1997. An architecture for extensible workflow process servers. Ph.D. Thesis, Columbia University Department of Computer Science. CUCS-014-96. ftp://ftp.psl.cs.columbia.edu/pub/psl/CUCS-014-96.ps.gz.

Tong, A.Z., Kaiser, G.E., and Popovich, S.S. 1994. A flexible rule-chaining engine for process-based software engineering. *9th Knowledge-Based Software Engineering Conference*, Monterey CA: IEEE Computer Society Press, pp. 79–88. ftp://ftp.psl.cs.columbia.edu/pub/psl/CUCS-005-94.ps.Z.

Valetto, G. and Kaiser, G.E. 1996. Enveloping sophisticated tools into process-centered environments. *Journal of Automated Software Engineering*, 3:309–345. ftp://ftp.psl.cs.columbia.edu/pub/psl/CUCS-022-95.ps.gz.

Wallnau, K., Long, F., and Earl, A. 1996. Toward a distributed, mediated architecture for workflow management. *NSF Workshop on Workflow and Process Automation in Information Systems: State-of-the-Art and Future Directions*, Athens GA. pp. 74–84. Position paper.

Weissenfels, J., Wodtke, D., Weikum, G., and Kotz-Dittrich, A. 1996. The Mentor architecture for enterprise-wide workflow management. *NSF Workshop on Workflow and Process Automation in Information Systems: State-of-the-Art and Future Directions*, Athens GA. pp. 69–73. Position paper.

Workflow Management Coalition Members. 1995. Coalition overview, http://www.aiai.ed.ac.uk/WfMC/overview.html.

Automated Software Engineering 5, 133–137 (1998)

A Guide to Rearing Perfect Software: A Desert Island Column

DOROTHY E. SETLIFF setliff@ee.pitt.edu
University of Pittsburgh, Pittsburgh PA 15261, USA

Near the end of my sabbatical year in Sydney Australia, I received e-mail from Bashar Nuseibeh, asking me to write a *desert island* column. This column asks what I would bring, or consider essential, if marooned on a deserted island. The book selection process turns out to be an intriguing search, sort, and knapsack problem. The search takes place in the domain of relevant books. Relevancy is a philosophical determination reflecting my personal history and my view of the future. Sorting uses relevancy to prioritize these books. The knapsack problem occurs with the notion of maximizing utility in the future within the editorial constraints. Ironically, I had gone through just this kind of activity when I packed for my sabbatical year. I was limited to what I could bring with me to Australia both by the ruinous cost of shipping and the airline travel allotment.

Before going into my list of important books, I would like to review my selection philosophy. My historical interests have two dimensions. The first dimension is in the design automation of large software systems. The second dimension is in the academic teaching of software skills to young university students. These two dimensions are surprisingly similar. An overwhelming characteristic in designing and implementing large software systems is the need to plan and get along within a set of well-defined rules. One such well defined 'rule', or process standard is the CMM (Paulk et al., 1993). These standards are guides to formulating a social order. Their bottom-line truisms, in their most simplistic forms, are 'think of others' and 'think before you do'. In academia, the overwhelming messages in curricula are these same truisms. Learning should not be a matter of mastering syntax, but rather, it should be a maturation process from a self-centered individualism to a more pluralistic, communicative methodology. My selection philosophy focuses on an inspirational, rather than rote, fulfillment of these two historical interests with an eye towards the future.

Selection No. 1: And with a drum roll

The first book and the top book in my selection process, talks about the importance of thinking of others and of human interaction. A *reasonable* human society can only exist when people think of the their own needs within the context of the needs of others.

> "*Manners are the folk customs of any given society. Those who violate them are almost certain to arouse extreme distaste in those who obey them.*"

Of course, this book is Judith Martin's *Miss Manners' Guide to Rearing Perfect Children* (Martin, 1984). The goal of this book is teaching parents how to raise well-mannered children. Well-mannered people are capable to living within the larger society of humanity. Here, society does not refer to snobbery, but rather, to the simple getting along with one's fellow man. Immaturity is synonymous with poor manners, while maturity is synonymous with civilization and good manners.

"Manners are the basis of civilized society, and passing on the civilization to the young, so that they do not run around in a natural and savage stage but can live easily and comfortably within the accumulated traditions and standards of their society, is what child-rearing is all about."

In a real sense, software engineering may be differentiated from simple programming by the need to get along with other entities, no matter if these other entities are other pieces of software, hardware, or complete systems. Software engineers must learn the concept of manners to be successful. Software engineers must know the folk customs; e.g., software process and organization standards, present in software engineering, and they must think of the needs and requirements of all stakeholders before and during all activities (e.g., design, programming, testing, etc.).

As an academic, I am involved with the teaching of software engineering to the 'young'. I view my efforts as attempts in fulfilling Miss Manners' directives above. A good educational environment should produce not just well-rounded graduates, but also, it should produce *mature* graduates. Maturity means thinking beyond one's own needs; i.e., to have manners. Maturity implies an understanding of the need and purpose of a software process or methodology. Ironically, some of the most capable students rebel against this approach, claiming any structured process impedes their creativity. In response, I quote Miss Manners', in discussing the recent societal urge to let immature beings grow without the structure given by manners, says:

"On the subject of manners for children, many adults believe that the opposite of 'polite' is 'creative'. No, the opposite of 'polite' is 'impolite'."

I believe education must focus on the not-so-delicate rearing of software engineers and apply Miss Manners' maxims in the classroom. Specifically, there must be an emphasis on process and maturity through out the curricula and a presentation of the benefits of thinking of other stakeholders during the software process. Syntax or pure theoretical information is not the important issue in raising mature students. Graduates must have experienced a mature, human-centered communication process.

Selection No. 2: History

However, rest assured dear reader, I do not believe that manners alone are the silver bullet guaranteeing an order-of-magnitude improvement in software productivity. However, manners can go a long way towards solving the basic programming-in-the-large problem:

people. The second book in my list was one of the first books to tackle the cost and result of these human issues that are so inherent when building large software systems. That book is *The Mythical Man-Month*, the 20th anniversary edition by Fred Brooks (1995). MM-M reviews the pratfalls and successes during the development of the OS/360 software way back in the "dinosaur" 1960's. The emphasis on the human issues, rather than on the technical issues, make this book timeless. MM-M begins by noting the cost and complexity differences between programming-in-the-small and programming-in-the-large. The social costs, i.e., the costs related to human communication and planning, are what doom the concept of the man-month.

> "*Men and months are interchangeable commodities only when a task can be partitioned among many workers with no communication between them.*"

> "*Since software construction is inherently a systems effort - an exercise in complex interrelationships - communication effort is great, and it quickly dominates the decrease in individual task time brought about by partitioning. Adding more men then lengthens, not shortens, the schedule.*"

The realization that people and management are the problem, not the software, is staggering and bears repetition by software engineers and their management early and often. While there are some anachronisms in this book, e.g., the insistence of building an initial system meant to be thrown away, the benefit of learning and understanding the historical experience is important in its own right. The majority of the book discusses morals and lessons learned during the process of developing the OS/360. It is interesting to read these morals and lessons within the context provided by Miss Manners'. Morals and lessons such as: document for others, not for yourself, specify external interfaces before internal implementations, an emphasis on comprehension, etc. This book provides a blow-by-blow application of the concepts described by Miss Manners' to the field of software engineering.

Selection No. 3: Organization take (1)

Organizing and then communicating the accumulated traditions and standards is one of the major problems facing software engineering. Passing these traditions and standards to our young requires effective communication. Effective communication requires the information be technology-independent. In addition, no one book or text can capture all the accumulated traditions and standards in software engineering. However, there are books focusing on selected abstraction levels, e.g., software architecture, implementation, etc. One such book is the third book in my list, *Software Architecture Perspectives of an Emerging Discipline* by Mary Shaw and David Garlan (1996). This book presents a taxonomy of software architecture styles. The software architecture level of design is the organization of a system into components, their communication and protocols, control structures, as well as the composition of and into design elements. Just as Fred Brooks in MM-M, these architecture styles are presented in a technology-independent fashion. Technology-independence ensures the contribution to the establishing software 'folk custom' is usable now and in the future.

This book provides a basis of communicating large-scale design issues present in software architecture. The basic contribution is a storage medium for the accumulated software architecture traditions. This book coherently presents these traditions as a set of definitions, patterns, and processes. Examples of real systems illustrate the taxonomy with an emphasis on how the taxonomy manages complexity. Especially interesting are the discussions relating to automating software architecture style selection. Implicit in this discussion is the assumption that the taxonomy provides necessary and sufficient information to automate design, i.e., the taxonomy becomes a process standard. The book finishes by reviewing the list of maturity characteristics a student learning from this book should exhibit.

The value of recognizing big architecture patterns or similarities and then providing reuse on a large scale is profound. Reinventing the wheel, the general opposite of reuse, is rather savage. Personally, I would rather reap all the knowledge and accumulated traditions at all abstraction levels, beginning with software architecture.

Selection No. 4: Organization take (2)

The final book on my list is of the same nature as Shaw and Garlan's book but with a change of abstraction level from software architecture to design implementation. This book reflects my own personal design paradigm, namely, the use of patterns to build software. *Design Patterns, Elements of Reusable Object-Oriented Software* by Gamma et al. (1995) introduces a wonderful paradigm both in the development of large software systems and in the academic teaching of software and process. A design pattern consists of a name, a description of what behavior is solved by this pattern, a description of the solution, and best of all, the consequences of using this pattern. This is essentially a list of a design pattern's social interactions. The design pattern set provides a design vocabulary. This vocabulary, because it is at a higher abstraction level than the implementation language, reduces system complexity. Concrete examples round out the contribution.

This book distills the wisdom and experiences of expert object-oriented software designers and makes this knowledge available for use for the novice. The information is technology-dependent both in the focus on object-oriented design and in the language-specific examples. However, this book successfully records the culture of the experienced in a usable form for the young. Consider the case where students are brought to maturity thinking in terms of identifying patterns, rather than hacking an implementation. Reuse takes place at the design vocabulary level. This book presents a social order in which to think about and design software systems.

Conclusions

The irony of bringing this selection of books, with their collective emphasis on manners, social interactions, and software organization tactics, to a deserted island (where ostensibly there are no other people, much less an internet connection) does not escape me. There has been much work addressing the human issues relating to software engineering. However, the common thread in successful works is not necessarily whether there is an increase in the productivity of the one, but rather, whether there is an increase in the productivity of the many. The books here present a methodology of accumulating tradition and providing a toolset for the future. Brooks presents traditions in terms of lessons learned, Shaw and Garland by a set of software architectures, and Gamma et al. by a set of design patterns. These books presents a structured paradigm that implicitly and explicitly encourages designers to think of the needs of others while forming a civilized software society. Following such a paradigm could help produce that elusive goal: perfect software.

References

Brooks, F.P., Jr. 1995. *The Mythical Man-Month, Essays on Software Engineering*. 20th anniversary edition, Reading, MA: Addison-Wesley Publishing.

Gamma, E., Helm, R., Johnson, R., and Vlissides, J. 1995. *Design Patterns, Elements of Reusable Object-Oriented Software*. Reading, MA: Addison-Wesley Publishing.

Martin, J. 1984. *Miss Manners' Guide to Rearing Perfect Children*. New York: Atheneum Publishers.

Paulk, M.C., Curtis, B., Chrissis, M.B., and Weber, C.V. 1993. Capability maturity model for software, Version 1.1, CMU/Software Engineering Institute.

Shaw, M. and Garlan, D. 1996. *Software Architecture Perspectives on an Emerging Discipline*. Englewood Cliffs, NJ: Prentice-Hall.

MIX
Papier aus verantwortungsvollen Quellen
Paper from responsible sources
FSC® C105338

If you have any concerns about our products,
you can contact us on
ProductSafety@springernature.com

In case Publisher is established outside the EU,
the EU authorized representative is:
Springer Nature Customer Service Center GmbH
Europaplatz 3, 69115 Heidelberg, Germany

Printed by Libri Plureos GmbH
in Hamburg, Germany